Irish Media

IRISH MEDIA
A CRITICAL HISTORY
Second Edition

John Horgan and Roddy Flynn

FOUR COURTS PRESS

Set in 11.5 on 13.5pt Centaur by
Mark Heslington Ltd, Scarborough, North Yorkshire
FOUR COURTS PRESS LTD
7 Malpas Street, Dublin 8, Ireland
www.fourcourtspress.ie
and in North America for
FOUR COURTS PRESS
c/o ISBS, 920 NE 58th Avenue, Suite 300, Portland, OR 97213

A catalogue record for this title
is available from the British Library.

ISBN 978-1-84682-654-2

Printed in England by
CPI Antony Rowe, Chippenham, Wilts.

Contents

Abbreviations

BAI	Broadcasting Authority of Ireland
BCI	Broadcasting Commission of Ireland
BT	British Telecom
CNI	Commission on the Newspaper Industry
ComReg	Commission for Energy Regulation
DD	Dáil Debates
DTT	Digital Terrestrial Television
IC	*Irish Catholic*
IE	*Irish Examiner*
II	*Irish Independent*
IN	*Irish News*
INA	Irish News Agency
IP	*Irish Press*
IT	*Irish Times*
MA	Military Archives
MRBI	Market Research Bureau of Ireland
NAI	National Archives of Ireland
NLI	National Library of Ireland
NNI	National Newspapers of Ireland (latterly Newsbrands)
ODTR	Office of the Director of Telecommunications Regulation
PCI/OPO	Press Council of Ireland/Office of the Press Ombudsman
RTÉ	Raidió Teilifís Éireann
SBP	*Sunday Business Post*
SD	Seanad Debates
SI	*Sunday Independent*
SRH	Scottish Radio Holdings
ST	*Sunday Times*
SW	*Sunday World*
UCD	University College Dublin Archives
USNA	United States National Archives
UTV	Ulster Television

Introduction

The origins and development of media in Ireland go back almost five centuries to the introduction of printing in Dublin, when the *Boke of common praier* was published in 1551, but the development of media (in the sense in which that term is being used in this book) dates from almost a century later, and two centuries after the invention of printing in Gutenberg.

This book examines the history of the print and broadcast media in both parts of the island. Its analysis is not merely chronological, but is grounded in the belief that the relationship between these media and the communities they serve is a complex and subtle one, symbiotic and mutually revelatory. The media inform social and political change, as well as reflecting it, and the political economy, as well as the history, of the media industry are peculiarly vulnerable to changes in the wider economic field, where fluctuating advertising trends, the emergence of new media forms, and changing patterns of consumption all pose continually new challenges.

Since the first publication of this book in 2001, the pace of change has accelerated dramatically, and has made the publication of a second, revised edition particularly timely.* The switch from analogue to digital, the growth of social media, the new dominance of mobile, the increasing dominance of aggregators like Google and Facebook, and accelerated patterns of mergers and acquisitions following the 2008 economic crash have all impacted dramatically on the media landscape (Bell and Taylor, 2017), and have raised questions about the future to which few, if any, analysts or media managers have yet managed to construct answers.

These developments have led to the addition of several substantial new chapters dealing with events in print, broadcasting and new media since the first edition. There is also a new preliminary chapter reviewing the origins of the media in pre-independence Ireland.

While much has changed, the media today still face some of the same questions and issues they faced eighty years ago. What role do the media play in the modern state? Will access to the media continue to be a function of social or economic power? What are the responsibilities of journalists and of media owners, and what are their rights? What are the most appropriate

---

* Versions of some of the earlier chapters of this book were first published by Routledge in John Horgan's *Irish media: a critical history* (2001), and revised text from these is included here, where relevant, with that publisher's permission.

mechanisms for regulation? To what extent will the future of the media be determined by technology and by the development of professional journalism, and to what extent by commercial or political considerations? How dramatically will new media alter the shape of the playing pitch and the duration of the match?

These and other questions will have to be answered in the light of the experiences of the past and the challenges of the present. This book, therefore, is not only an essential backdrop for all these discussions, but also sets out important parameters within which future developments can be assessed.

1 Setting the scene: pre-independence Irish media

It is but just that the People should have News for their Money
— Irish Monthly Mercury, 21 Dec. 1649 (Morash, 2010: 22)

The first defining characteristic of the early media in Ireland was, effectively, geography. Ireland was not just on the periphery of Europe, but an extremely remote part of that periphery, in an age in which transport was primitive and slow. Communities were small and much essential information probably travelled by word of mouth, except for some record-keeping, particularly of an administrative or commercial nature. (Daily shipping lists of vessels and their contents, which would have been common in the coffee houses of the seventeenth century, were still being printed in Irish newspapers, North and South, in the early part of the twentieth.) What was news, then and now, was primarily what people didn't know, and high on the list of what people didn't know was what was happening elsewhere. This information was particularly important to them when what was happening elsewhere could affect commercial or political activity closer to home. It was also, of course, tightly controlled, and in English: the political administrative needs of the colonizers made sure of that. Not long after the battle of the Boyne in 1690 the public was informed of a new publication, *The Dublin Gazette*, the second of its kind to be licensed, 'in order that the Publick may not be imposed upon by any False Account of News', and was further reassured that the lords justices 'have appointed their Secretary to Peruse the same constantly before it is Printed' (Morash, 2010: 32–3).

Ireland, possibly because of its remoteness or the small scale of commercial publishing activity, was not covered by the English copyright legislation of 1709. The commonest form of news-related material was the newsletter, often imported from London or further afield (Munter, 1967); as printing became more popular, local news took second place, for the most part, to the production of books, pamphlets, and – of course – religious tracts and devotional material. Nonetheless, there were some early indications that – regardless of their geographical origins or political allegiances – a number of Irish opinion-formers and taste-setters were conscious, even in this early period, of the cultural, and perhaps also the political, implications of colonial or semi-colonial status. The editor of the *Dublin Weekly Journal* complained in his first edition in 1725 about the limitations imposed by provincialism on public taste, in that material emanating from London was accorded too much authority because of its origin, and 'Thus, our Brains are being manufactured abroad, become an Expence to the Nation; and we are forced to make a Purchase of our own Wit and Learning, which are hereby made hurtful to the Native Soil' (Munter, 1967: 160; Morash, 2010: 44). It is also noteworthy that the two oldest newspapers in continuous publication in Ireland today were not established in Dublin or Belfast (where the political and social currents ran more swiftly and changed course more often), but in provincial centres that would also have sustained important garrisons during the centuries of British rule: the *Limerick Chronicle* was founded in 1766; the *Derry Journal*, founded 1772, was the first newspaper in these islands to report the American Revolution, after a ship carrying the news across the Atlantic made Derry its first port of call.

There is also a strong line of descent from the remote past. Charles Lucas, a Dublin politician whose opposition to the exploitative and demeaning aspects of British rule made him a favourite of the *Dublin Weekly Journal* (and led the authorities to banish him from Dublin), was to reappear as the founder of one of Ireland's most famous newspapers, the *Freeman's Journal*, which lasted from 1763 to 1924 (Morash, 2010: 46).

Christopher Morash has pointed to the importance of improved transport links in spreading journalism throughout the country (2010: 48–9), but it is difficult, at this distance, to compare the growth of journalism in Ireland with, for example, what was taking place on the other side of the Irish Sea. That growth would in any case have been restricted by language and literacy levels, and the effect that these constraints had on the development of a market. The reverse of that coin is that the increasing number of commercially viable and English-speaking garrison towns helped to provide a readership for the products of newspaper proprietors or owner/editors who were (as well as being jobbing printers in most cases) also staunch upholders of the civil power.

This civil (and of course military) tutelage rested on insecure foundations, and, as the latter decades of the eighteenth century rolled with increasing momentum into the era of the American War of Independence and then the French Revolution, a distinct sense of nationhood began also to appear in Ireland, and to find its expression in new media like the *Volunteer Journal* (1782) and the *Volunteer Evening Post* (1783). Their approach ranged from the cautiously approbatory to the inflammatory, and eventually led to the political suppression of these two titles. The *Northern Star* (1792) was destroyed by a loyalist mob in the wake of the panic caused by the French attempt in 1796 to land an expeditionary force in Bantry Bay (Inglis, 1954: 111). The failed Irish rebellion of 1798 led not only to political convulsions but also to the passage of the Act of Union, and the disappearance of any substantial newspaper opposition to the crown and to its Irish presence in Dublin and elsewhere (LANI, 1987).

The fire was, however, still alight under the pot, and it was not long before new jets of steam began to emerge. Catholic emancipation, and the political rise of Daniel O'Connell as its champion, created an appetite for a press that would reflect this and other concerns of the political majority on the island, and helped to supply publications to meet this need. While papers like the *Dublin Evening Mail* (1823) and then the *Irish Times* (1859) remained the standard-bearers of Irish unionism in the South, and the *Northern Whig* (1824) and the *Belfast News Letter* (1737) fulfilled a similar function in the North, the emergence of the *Morning Register* in Dublin (1824) and the *Belfast Morning News* (1855), which in time became the *Irish News*, marked distinctions that were both confessional and political.

O'Connell used the Catholic press, notably the *Morning Register* and the *Pilot* (1828), adroitly to further his political aims, although the *Morning Register* – which was the first Irish newspaper to maintain a corps of dedicated reporters – was given to occasional manifestations of editorial independence. The achievement of Catholic emancipation in 1829 paradoxically took the steam out of his next campaign – for repeal of the Act of Union. That campaign, however, was not dead but sleeping, and was successfully revived with the launch of the *Nation* in 1842.

The *Nation* was among a number of periodicals (Hayley and McKay, 1987) that benefited from an extraordinary confluence of individuals, trends and more long-term developments. The individuals were its three editors, primarily Thomas Davis, whose journalistic and literary skills were extraordinary. The trends were a renascent nationalism and improvements in communications systems generally, and the more long-term developments were the spread of literacy in English through the national schools system and the creation of repeal 'clubs' in which the paper was read aloud to listeners who could not afford to buy it.

Events of a different kind, however, conspired to reduce the flame lit by the *Nation* to a low glimmer. These were the Great Famine of 1845–7, the death of O'Connell in 1847, and the split between constitutional and revolutionary nationalists that was reflected in the abortive rebellion of 1848, the year in which the *Nation* was suppressed, although it reappeared the following year in a different guise.

The press in Ireland demonstrated little vitality for some decades thereafter, but it secured a new lease of life with the emergence of the Land War, a period of agricultural agitation exacerbated by a produce-price slump in 1874. This political agitation led, in turn, to the establishment in many market towns, especially in the west of Ireland, of weekly newspapers supporting tenants' rights, and the strengthening of some that had been established earlier. These included, among many others, the *Tuam Herald* (1837), *Leinster Leader* (1880), *Kilkenny People* (1893), *Southern Star* (1889), *Leitrim Observer* (1889), *Connaught Telegraph* (1828), *Sligo Champion* (1836), *Waterford News and Star* (1848) and the profitable and influential Munster paper the *Cork Examiner* (now the *Irish Examiner*) (1841).

As the Land War developed into something more militant, and the development of the Fenian movement among Irish emigrants in the United States sharpened the tone of Irish resistance to British administration, Fenian adherents began to crop up in journalism. In the late 1860s, editors and newspaper correspondents (along with teachers) were frequently suspected of Fenian membership or sympathies (Legge, 1999: 138). As the nineteenth century ended, many of the older local newspapers that had been set up to reflect the unionist elite began to falter, like the *West Cork Eagle* (1857–1928), which was overtaken and eventually assimilated by the *Southern Star.* By 1900 there were very few of these small unionist papers left: 'between 1878 and 1890, 24 papers changed their political allegiance to nationalism; to these must be added the twenty nationalist papers founded during this period'. (Legge, 1999: 128).

It should not be forgotten that, throughout this later period, the print media in both Ireland and England shared, to varying degrees, an information ecology. British newspapers, which had always circulated in Ireland to a limited extent, became even more popular in the early years of the twentieth century, and up to the Treaty. There was also a flexible marketplace for Irish journalists, many of whom found it as easy (and sometimes easier) to find profitable work in London (Sheehy, 2003). Irish fissiparous tendencies were as obvious in Britain as in Ireland. One group of journalist émigrés were nationalists, who, to a considerable degree, integrated themselves into British political and cultural life. T.P. O'Connor, Justin McCarthy and R. Barry O'Brien were the exemplars of this tendency. A parallel strand, however, particularly from the early 1890s onwards, involved people like W.P. Ryan and

D.P. Moran, who were influenced more by the Irish cultural and ultimately political revival than their forerunners, and were among those who returned to Ireland in the early 1900s to engage with what was happening there.

There were rumblings of a revolution in the Irish print media that predated, and to some degree influenced, the 1916 Rising, but the casual visitor to Dublin in the first decade of the twentieth century would have noticed little out of the ordinary. Two royal visits had taken place in that period – Queen Victoria's in 1900 (her fourth), King Edward VII's in 1903 – and King George V visited shortly afterwards, in 1911.

The mainstream nationalist press, including the *Irish Independent*, which had been founded in 1905 and had a circulation of more than 50,000, and the *Freeman's Journal*, which had a circulation of approximately 30,000 (O'Brien and Rafter, 2012; Larkin, 2013), were suitably deferential on these occasions. However, the failure of the British government to legislate for home rule in 1913, followed by the passage of home-rule legislation in 1914 that accepted the political inevitability of partition, seriously undermined these newspapers' support for John Redmond's Irish Parliamentary Party at Westminster. In the North, however, the nationalist *Irish News*, with its circulation of about 30,000, remained closer to the Irish Parliamentary Party because of the influence of the Northern nationalist politician Joe Devlin, who was in 1920 to become chairman of the company that published the paper.

The unionist *Irish Times*, which had a circulation of about 25,000 (Brown, 2015), opposed home rule not only because it threatened the link with empire, but because partition would cut off Southern unionists in what later became the Free State from their fellow unionists (and, frequently, co-religionists) in what became Northern Ireland. Dublin's *Daily Express* (1851–1921) represented a more raucous form of unionism, which engendered quite substantial support.

In the North, the unionist *Belfast News Letter*, with a circulation of about 30,000, and the nationalist *Irish News*, with a circulation of about the same size, mirrored the political views of their readership – unionist and nationalist, respectively – very closely. The *Belfast Telegraph*, which had been founded in 1886, and had a circulation of about 40,000, protected its monopoly on the lucrative evening-newspaper market by adopting a more soft-focus unionism. The more extreme unionist *Northern Whig* (1824–1963) sold only around 15,000 copies.

The stirrings in the political undergrowth were more evident to readers of the advanced nationalist and socialist publications that had, since before 1900, become increasingly critical of the constitutional status quo, particularly in the capital. A raft of these, of which the more significant were *Irish Freedom, Sinn Féin* and the *Irish Worker* (a socialist paper that probably at its height, during periods of intense industrial unrest, reached an astounding circulation of around 40,000), were among the first suppressed under the Defence of the Realm Act in December 1914. From then until the Treaty in 1921, the British authorities and the various editors and journalists involved in these papers played a cat-and-mouse game punctuated by many changes of title, arrests and seizures. During the Rising itself, the rebels distributed a small newspaper called the *Irish War News*, most of which appeared to have been pre-written, and which never achieved more than a curiosity value.

In the capital, the newspaper forces were more or less evenly divided. The *Irish Times* and the *Daily Express* (which did not long survive) were unionist. The *Freeman's Journal* was wholeheartedly Redmondite (the Irish Parliamentary Party supported it financially, as it fought against the rising tide of William Martin Murphy's *Irish Independent*). The *Irish Independent*, for its part, espoused an idiosyncratic nationalism which was tempered by a deepening suspicion of Redmond and the IPP, particularly on the questions of corruption, partition and the North. Murphy privately described the IPP members at Westminster in a letter to his editor, T.R. Harrington, on 8 June 1916, as 'jobsters' (NLI: Harrington Papers).

To many observers in Britain and elsewhere, the 1916 Rising came as a complete shock. Irish political concerns, it had seemed, had been largely pacified by the end of the nineteenth century, and the island, while not as docile as Wales or Scotland, seemed integrated, for the most part, into a United Kingdom whose politics, mores and media were those of the Edwardian era.

More perceptive observers than many of those in Dublin Castle or the dominions office in Whitehall would, however, have been able to spot bubbles of change bursting on the surface of what was otherwise a quiet, even stagnant political scene. Nowhere was this more evident than in the media. Since the late 1890s, in Belfast as well as in Dublin, a rash of small period-icals had been appearing. Some were political, some literary, some a mixture of both. What they had in common was the conviction that good government was no substitute for self-government. Some of them were edited by men who believed that argument should be accompanied by organization. Some of them, supported by political subventions, were a vital part of the secret work of rebellion.

The world they inhabited was utterly foreign to that in which the mainstream media held sway. The Rising had a tsunami-like effect on the press. Although unionist papers generally contended (without offering anything convincing by way of evidence) that it had been foreseeable for ages, and that Augustine Birrell's administration in Dublin Castle should have prevented it, the nationalist papers were evidently caught totally unawares. A stone's throw from the GPO, the *Irish Independent* was appraised of developments almost immediately. As Maurice Linnane, an *Independent* reporter, remembered many years afterwards:

> One of our caretakers, a man with plenty of avoirdupois named Pat Gaffney, shoved in the door of the reporters' room. 'The Volunteers have gone into the GPO. They are breaking the glass with the butts of their rifles and sandbagging the windows.' My colleagues and I rushed wildly down the stairs and into the street. (*II*, 11 Apr. 1995)

Just across the Liffey, the atmosphere at the *Irish Times* was more restrained, even languid:

> Few people heard the beginnings of the official declaration of an Irish Republic. Fewer stayed to the end ... A revolutionary went to Nelson's Pillar to make his speech which gained fervour and thundered out the phrases he had used so often, before his audience became progressively bored. (*IT*, 25 Apr. 1916)

All the Dublin newspapers suffered breaks in production because of the fighting. The *Irish Times* was the first to reappear, and, in its editorial on 1 May, it unforgettably demanded:

> The State has struck, but its work has not finished. The surgeon's knife has been put to the corruption of the body in Ireland, and its course must not be stayed until the whole malignant growth has been removed.

The *Freeman's Journal*, which reappeared on 5 May, spoke of the 'stunning horror of the past ten days', and argued that 'the insurrection was not more an insurrection against the connection with the empire than it was an armed assault against the will and decision of the Irish nation itself, constitutionally ascertained through its proper representatives'. It then turned its fire on the

Irish Times, charging that its 'malignant growth' editorial was a 'bloodthirsty incitement to the Government'. The *Irish Times* responded a day later:

> Our readers know that all this is wicked nonsense. We have called for the severest punishment of the leaders ... It would be the worst kind of folly and the poorest sort of economy to shear the stalk of sedition and leave the roots uninjured.

The two editorials that the *Irish Independent* published on 10 and 12 May have generally been interpreted as calling for the execution of James Connolly, and Murphy is credited with having motivated them, even if he did not actually write them, because of his experiences with Connolly and Larkin three years earlier. However, this was challenged by contemporaries, particularly by the anti-Parnellite MP Tim Healy, when writing to his brother in October 1916. In Frank Callanan's biography of Healy, Murphy is depicted as a man thrown off balance by the tumultuous events he has just lived through, unwilling or unable to use the power of his newspapers for political ends (or at least those ends desired by Healy), and deeply unhappy at the fact that he was being held responsible for the executions. Murphy, according to his own version of events, as told to Healy:

> was not responsible and did not know of the articles recommending 'vigour' and that they were written by Harrington, and, until his attention was called to them, he had not even read them. He was evidently greatly affected by the thought that he had been accused of advising the shooting of Connolly, owing to the antagonism the man showed him. He said of course at first he felt bitter against the insurgents but, finding all the Tories gloating over the executions and imprisonments, he said every drop of Irish blood in his veins surged up, and he began, like others, to sympathise with them. He was quite moved; and his face flushed. (Callanan, 1996: 517)

In the North, the *Irish News*, owned and controlled by nominees of the Catholic bishops, gave the *Independent* and the *Irish Times* both barrels. Arguing that the real rebellion was against the Irish Parliamentary Party, it charged that 'the hand of the *Independent* was on the handle of every pickaxe wielded against the foundations' (*IN*, 5 May 2016). And the following day it proclaimed that there was not a decent man in the country:

who will not shudder with repulsion at the spectacle of that infamous incarnation of hypocrisy, the *Irish Times*, sitting on its uninjured perch in Westmoreland Street, and shrieking aloud for blood and more blood, like a monstrous combination of witch and vampire … The *Irish Times* helped to manure the 'malignant growth' when it assured the Volunteers that manliness and true patriotism should induce them to follow the MacNeills and the Casements. Now it brazenly advocates the wholesale slaughter of those who took its advice. (*IN*, 6 May 1916)

The ire of editor of the *Irish News* may have been intensified as he realized that the *Irish Times* had borrowed its 'surgeon's knife' metaphor from an editorial in the *Irish News* on 3 May that castigated the rebels for the wholesale destruction they were wreaking, not only on Dublin, but on the Irish Parliamentary Party. Dublin was eventually rebuilt: the IPP was doomed.

As far as newspaper editorial policies were involved, the emphases were clear in the run-up to the 1918 general election. Unionist papers in the North, in particular, were concerned about the possibility of a split in the unionist vote, particularly emanating from more left-wing unionists, and from the nascent Labour Party.

In the South, the *Irish Times* advised husbands that if their wives pleaded domestic engagements as an excuse for not voting, they should tell them 'that on this occasion his dinner is less vital than her vote' (*IT*, 14 Dec. 1918). The principal nationalist newspaper, the *Irish Independent*, was blinkered by what it saw as the Irish Parliamentary Party's betrayal of nationalism by its acceptance of partition, and concluded morosely:

It is clear that Ireland has nothing to hope for from the Parliament which is now elected. Even were her representatives to attend in their full strength they could make no impression on an assembly impervious to an appeal for justice, and predominantly hostile to any application of the principle of self-determination to Ireland. (*II*, 19 July 1917)

Post-election there were troubling signs and portents: nationalist newspapers were censored and printing plants destroyed (Walsh, 2015: 27). The summoning of the First Dáil on 21 January 1919 marked a new phase, accompanied as it was by the shooting, on the same day, of two policemen in Tipperary. The Dáil's media-related activities moved effortlessly up through the gears under the tutelage of its head of propaganda, Erskine Childers, and

others. A growing interest from foreign journalists in Irish affairs was spurred
on by extraordinary events such as the Alcock and Brown landing in June 1919.

The *Irish Bulletin*, which appeared for the first time on 11 November 1919, was
first edited by Desmond FitzGerald (father of a future Irish prime minister,
Garret FitzGerald), and later by Childers. Aided by experienced Irish
journalists such as the Cork-born Frank Gallagher, this paper, printed and
published clandestinely, displayed many of the primary characteristics of
independent journalism – verification of names, dates, places and events – and
was circulated assiduously to foreign journalists, prompting many of them to
visit Ireland and report on events here. Other Irish papers suffered heavily from
censorship, both directly and because – even if they escaped direct censorship
– their columns were open to governmental attack under the Defence of the
Realm Act regulations. The chief object of both was not just to limit, but where
possible to suppress, any mention of radical nationalism and its varied manifestations. Although direct censorship ended in September 1919, the DORA
regulations continued in force, evidenced by 'a wave of newspaper suppres-
sions' (Kenneally, 2008: 7), including of the *Cork Examiner,* the *Freeman's Journal*
and numerous provincial newspapers. The administration even went so far as
to publish its own version of many of the events which occurred during the War
of Independence – the *Weekly Summary* – which was treated with suspicion and
occasionally outright disdain, such as when it disingenuously contributed to
the partial suppression of the Strickland Report which had unambiguously
blamed crown forces for the burning of large parts of Cork city on Christmas
Eve 1920 (Kenneally, 2008: 25).

As political uncertainty morphed into the armed campaign of militant
nationalism, the role of foreign newspapers such as *The Times,* the *Manchester
Guardian,* the *Daily Mail* and the *Daily Mirror* (the last two of which would have
had substantial circulations in Ireland) was also significant. For many of these
journalists, arriving in Ireland allowed them to experience the media-handling
expertise of Childers and his team in a way designed to win not just journalistic but international sympathy for the cause (Kenneally, 2008). The growing
influence of Sinn Féin on civil matters was also the subject of newspaper
comment, much of it exasperated, as when the *Irish Times* observed that 'the
Sinn Féin tribunals', courts set up by the party for the resolution of disputes
and the prosecution of minor criminal offences, were 'jostling British law into
oblivion' (Walsh, 2015: 144).

With the approbation of the Treaty by the Dáil in 1921, and the creation
of the Free State, this dire prophecy was to a large extent fulfilled.

2 The new order, 1922–32

The establishment of the Irish Free State in January 1922 may have marked the beginning of the dissolution of the British empire, but this was as yet far from apparent, and in Dublin such global considerations were far from people's minds. In every sphere of public and private life, the business of adjustment began apace. The media were not exempt; indeed, they were as unprepared as any sector in society for the sea-change that was in progress, and in some cases had rapidly to adopt, chameleon-like, a range of new editorial positions as they grappled with emerging political and social realities.

The media landscape in 1922 had been disturbed, but not fundamentally altered, by the trauma of the War of Independence, which had effectively begun in 1919. The advanced nationalist press had, to a considerable extent, disappeared under the censorship legislation – notably the Defence of the Realm Act – after the 1916 Rising, and, although a number of subterranean or samizdat-type publications like the *Irish Bulletin* had become extremely efficient propaganda outlets for the growing Irish republican military and political forces, mainstream media were still by and large conservative: they supported home rule rather than republicanism, or unionism rather than home rule. During the most intense period of the War of Independence, they co-operated for the most part willingly with the office of the censor, so much so that when he finally relinquished office at the time of the Treaty he was moved to write to the editors expressing his appreciation at the degree of understanding they had shown for his task.

Prior to the Treaty, the existence of the United Kingdom of Great Britain and Ireland was reflected in the nature of the media available in the smaller island, and in the balance between British and Irish titles. At this remove in time, no circulation figures can be quoted authoritatively, but circumstantial evidence suggests that the major British newspaper titles had substantial circulations in Dublin, Belfast and elsewhere on the island. Adherents of the main religious groups were differentially distributed across the island. In the six counties that were to become Northern Ireland, the Protestant denominations formed a comfortable majority. In the twenty-six counties of what became the Irish Free State (later the Republic), they formed a significant minority.

The *Irish Times*, whose circulation at this time was static, or worse, at around

20,000 copies a day, was predominantly the paper of the Protestant commercial and professional classes. Protestant landowners and the titled gentry would have read, instead or additionally, *The Times*, the *Daily Telegraph* or the *Daily Mail*. Lower-middle- and working-class Irish Protestants would have read the more popular British titles. Protestants even had an evening paper to reflect their special interests: the *Dublin Evening Mail*, founded in 1823 (it was to drop the word 'Dublin' from its title in 1928). Few of them would have read any of the Independent group's titles – the *Irish Independent, Sunday Independent* or *Evening Herald*. This group of papers appealed primarily to the growing Irish Catholic middle classes and to the farming community. Since 1905, when they had been relaunched, they had become the first indigenous newspapers whose circulation was effectively national. They did not have the field entirely to themselves: the *Freeman's Journal*, which had a chequered history going back to its foundation in 1763, and whose by now vestigial anti-clericalism attracted the lingering admiration of James Joyce, had suffered enormously from supporting Parnell after his divorce, and was experiencing a terminal slide in circulation. In Cork, the principal papers were the *Cork Examiner* and its sister paper the *Evening Echo*, under the control of the Crosbie family. After the collapse in 1924 of its unionist rival the *Cork Constitution* (founded in 1822), the *Cork Examiner* and the *Evening Echo* were the only daily newspapers published outside of Dublin and Belfast.

Outside the major urban centres, a whole range of weekly provincial papers survived and even prospered. It was a journalistic landscape that had evolved dramatically in the second half of the nineteenth century, as the old network of small, Protestant-owned papers, situated for the most part in garrison towns, was supplemented, challenged and in some cases obliterated by the growth of nationalist papers. Only in Northern Ireland was the traditional bifurcation maintained, with both nationalist and unionist local papers continuing to flourish, some of them appearing twice or even three times a week.

Ireland had as yet no indigenous broadcasting. The BBC, however, was widely available and listened to. Its schedules were published as a matter of course in all the daily newspapers, and its programmes were the subject of weekly commentaries by radio critics.

As the new government took power, an accelerated period of readjustment ensued. The *Freeman's Journal* was, of all the national daily papers, the most enthusiastic supporter of the administration. The *Irish Independent*, long a supporter of home rule (although a visceral opponent of partition), found the transition to de facto independence relatively easy to make. Even the *Irish Times* discovered unexpected reserves of enthusiasm when it editorialized:

> No Irishman can watch without pride and hope the beginnings of our native Parliament … the temper of this assembly of young and untrained men, its practical quality, its skill in intricate debate are one of the happiest omens for the country's future. (*IT*, 7 Oct. 1922)

The welcome would probably have been more grudging but for the fact that Eamon de Valera, the sole surviving commandant of the Rising, had together with many who had fought in the War of Independence, rejected the Treaty, and walked out of the Dáil after they were on the losing side in the vote to ratify it. A government led by W.T. Cosgrave and manned by the pro-Treaty majority was a different proposition from one that might have included de Valera himself and firebrands like Liam Mellows and Cathal Brugha. Not only were these men determined on a total break with Britain, some of them, like Mellows, were infected by an even worse disease: socialism. Their exclusion not only from government but – by their own action – from the Dáil itself, not only put a premium on supporting Cosgrave, but made it a more congenial task.

The learning curve, for journalists and politicians alike, was to be a steep one. On 14 April 1922 the dissident republicans occupied the Four Courts, the centre of the judicial system and the repository for the public records of Ireland going back to the fourteenth century. After a few months of stalemate, the Civil War broke out in earnest, and was to last almost until the end of 1923.

The conflict was not only military. There was a plethora of newspapers published by or on behalf of each side. The government published *An Saorstat* (*'The Free State'*) from February to November 1922. It was only notionally a newspaper, in that its columns were mostly filled by commentaries on current events, and by appeals to specific groups for political support (e.g. 'Women and the Treaty', 18 Mar.). These commentaries alternated with fairly ham-fisted attempts at political satire; the article 'Revised definitions – a Dictionary for the Times' noted, under 'de Valera', 'See "Dictator"' (11 Apr.). The infrequent advertisements in its pages, when they were not for political meetings to be addressed by pro-Treaty leaders, were for expensive consumer items such as motorcars and fur coats, indicating strong support for the embattled government from the upper classes. Government supporters P.S.

O'Hegarty and Sean Milroy published similar papers – entitled the *Separatist* and the *United Irishman*, respectively – in 1922 and 1923. Ministers even briefly considered, before rejecting, a scheme by a military officer to start a new pro-government paper in the west of Ireland, an area in which pro-government sympathies were noticeably absent. The government accepted the suggestion of its press adviser, Sean Lester, a former news editor of the *Freeman's Journal*, to ban public advertising from provincial weekly newspapers sympathetic to the republican cause (Horgan, 1984b: 55).

The republicans had their own papers, including *The Fenian*, a semi-underground paper that appeared for the first time in July 1922 and continued intermittently until October of that year. It contained bulletins on the activities and occasional successes of the anti-Treaty forces, and used on its masthead the dateline 'The 7th year of the Republic', in place of a more conventional chronology. Its articles were, for obvious reasons, all unsigned, and habitually referred to the 'Slave State' rather than the Free State. *Éire* and *Poblacht na hEireann*, which was suppressed in December 1922, were similar publications. Labour published its own sectional paper, the *Voice of Labour*, for the first time also in 1922 (it lasted until 1927). In this welter of short-lived political journalism, the foundation of *Dublin Opinion* in 1924 stood out: a magazine specializing in light humour, it had a remarkable existence until 1968.

Government and republicans alike realized early on that newspapers were as important as any territory being fought over, and both sides evolved media-management techniques that varied from persuasive to intimidatory. Most mainstream newspapers, for their part, while generally supportive of government, felt it necessary from time to time to defend their independence, to an extent that ministers sometimes found little short of treasonable.

The government set up censorship mechanisms in a hurry. One of the chief issues was the way in which the anti-Treaty forces would be described. As far as the government was concerned, any description that gave the impression that the struggle was between two equally legitimate forces was anathema: so the publication of the military ranks used by the anti-Treaty forces was banned, as was the use of the words 'forces', 'troops' and 'army' to describe their combatants. Worst of all was the term 'republican'. The members of the Cosgrave government, although they had been forced to accept a political solution that fell short of the hoped-for republic, still regarded themselves as ideological republicans, and did not take kindly to their opponents' appropriation of the word.

Even as the hostilities intensified, there were differences of opinion within the government. Michael Collins, who was chiefly involved with prosecuting the war, wrote to one of his colleagues on 26 July:

> The censorship was too strict for my liking. It is unquestionably a fact that the average reader of the newspaper discounts to an undue extent censored news, and in any case we have the situation so well in hand now from the public and military point of view that the newspapers themselves may be trusted to do what is right if only they are spoken to occasionally. If there are rare instances of departure from the spirit in which we allow freedom of the press, then I think we should take such cases up separately. (NAI, Taoiseach: S 1394)

On 20 October, some two months after Collins' death, another government military leader, General Richard Mulcahy, called in all the newspaper editors to urge them to convey the impression that the fighting was over by suppressing or minimizing the reports of activities by the 'Irregulars', as the government preferred their opponents to be described.

This was a response to the fact that the government's censorship policy was not operating in a vacuum. The republican forces were engaged in widespread intimidation of newspaper proprietors and editors in areas of the country (principally the south, south-west and west) over which they exercised a large measure of control. The *Kerry People* in Tralee and the *Nationalist* in Clonmel, Co. Tipperary, were two of those successfully intimidated. In Cork, where the entire city was for a time under the control of republican forces, the *Cork Examiner* and its evening paper were forced at gunpoint to carry the full text of republican press announcements.

Nor were their activities confined to provincial newspapers. Captured documents from the period, in the Military Archives (MA: A, 10657; US, 10), suggest strongly that republicans directly intervened with the managements of both the *Irish Times* and the *Irish Independent* in an attempt to secure more balanced coverage, and that these interventions were to some extent successful. It was not so critical for the *Irish Times*, whose core circulation was in Dublin; but the *Irish Independent*, which had a large proportion of its sales in non-metropolitan areas, was susceptible to a boycott that apparently lasted for two weeks. Although the paper vigorously denied that it had come to any understanding with the government's opponents, the government censors believed that it was 'insidiously, rather than openly, doing its best for the

Irregulars ... by the mere working of a passage to imply that this is not a revolt against a constituted government ... but a fight between two factions' (NAI, Taoiseach: S 1394).

The *Freeman's Journal* did not disguise its loyalty to the government, and suffered in consequence. When it published in March 1922 accurate but uncomplimentary information about the inner workings of the republican movement, which had been deliberately leaked to it by the authorities, armed republicans entered its city-centre premises in Dublin and destroyed the machinery with sledge-hammers. Newsagents in counties Sligo and Mayo were terrorized into refusing to stock the paper, and the irregulars then adopted a more labour-effective method by intercepting the mail train at Limerick junction and burning all copies of the *Journal* destined for counties Limerick, Cork, Tipperary, Waterford and Clare. The headquarters of the IRA, as the republican forces described themselves, issued instructions to their local units to expel hostile journalists from their areas (NAI, Justice: H5/55).

As hostilities dwindled, leaving only a few pockets of resistance, the government turned its attention to other threats, notably from foreign correspondents based in Ireland. When the army complained about the tone of articles appearing in the London *Daily Mail* in February 1923, the government seized all copies as they were being unloaded from the mail boat in Dún Laoghaire, and suppressed it for four days. The *London Morning Post* fared even worse: its correspondent in Dublin, T.A. Bretherton, had already suffered the indignity of being horsewhipped on the steps of the Kildare Street Club by the nationalist Conor O'Brien (O'Brien, 1998: 38) for comparing the Catholic Irish to monkeys. He now enraged the authorities with a series of articles in March and April 1923 in which he maintained, among other things, that:

> the Irish Free State has no assets, is spendthrift, is already heavily in debt, and has in its midst a large and, it is to be feared, growing section of the people who deny the existing government's claim to represent the nation or sign its notes of hand. (Horgan, 1984b: 52)

The government's director of publicity, Sean Lester, recommended that Bretherton be either imprisoned or deported. Wiser counsel prevailed, however, and the government confined itself to making diplomatic representations about him. Military censorship withered and died after the *Irish Times* successfully defied, in July 1924, a government instruction not to print, in its paid death notices, the military ranks ascribed to republicans.

The *Freeman's Journal* was a permanent casualty of the hostilities. Although it had resumed publication after the destruction of its presses by the republicans, the growth of the *Irish Independent* had left it completely without a market niche. It moved briefly to sensationalism in an attempt to boost sales, but this backfired when the police, stirred into action by a strident campaign it was waging against the city's numerous brothels, raided one of them, only to discover a senior journalist from the paper enjoying its facilities (Oram, 1983: 155). Now that the Civil War was over, de Valera and his allies were considering the possibility of taking over or starting a daily newspaper to counter the overwhelmingly anti-republican ethos of all the other dailies. They set their sights on the *Freeman's Journal.* Despite its fading fortunes, it represented an attractive option, partly because it had an existing readership (however small), but more so because of its tradition of independence in the late nineteenth century.

The republicans nearly succeeded – indeed, Lester believed that they had all but managed to amass the purchase price – but then the Murphy family, proprietors of the *Irish Independent,* stepped in as the *Journal* ceased publication in December 1924 and bought paper and premises for £24,000. Conspiracy theorists might hold that this was done at the instigation of the government; but the more likely explanation is that it was a purely commercial move designed to stifle the possibility of any opposition in the marketplace.

In 1926, elements in the defeated republican forces – still without a newspaper of their own – formed a new political party under de Valera, called it Fianna Fáil, and began their tortuous journey towards full participation in democratic politics. In 1927, de Valera made one of a number of visits to the United States that combined fundraising for his new organization with a detailed study of newspaper-ownership models and management practices.

His impatience on the subject of starting a newspaper was particularly evident when one of his political opponents, Ernest Blythe, made a speech in 1929 suggesting that Ireland was happy to remain in the British commonwealth. He told one of his correspondents:

> If we had a daily paper at this moment I believe that Blythe's statement could be used to waken up the nation, but the daily press that we have slurs it over and pretends that nothing vital has been said. The English press of course are broadcasting it wherever they can. This is natural enough, for it is Britain's final victory over what remained of the Collins mentality and policy. (Longford, 1970: 270)

This development marked a decisive split in the anti-Treaty forces. Many of Fianna Fáil's former comrades maintained that this was a betrayal of principle and remained on the margins, a potent source of political and social tension. Their military organization, the so-called Irish Republican Army, was proscribed. IRA publications had been outlawed by the Public Safety Act of 1927, Section 9 of which made it a crime to publish anything 'aiding or abetting or calculated to aid and abet an unlawful organisation'.

More censorship was also in the air, but this time of the social or moral rather than the military or political variety. Article 9 of the constitution of the Irish Free State had laid down, in a somewhat negative context, that 'the right of free expression of opinion as well as the right to assemble peaceably and without arms, and to form associations and unions is guaranteed for purposes not opposed to public morality'. Already in 1923 the new government had passed, with barely a ripple of dissent, legislation to censor films. This was a powerful expression of the work of a number of important interest groups, notably the Vigilance Association, which had been set up by a group of Catholic priests and laymen in 1911 to safeguard Irish public morality, particularly against the incoming tide of foreign newspapers, magazines and motion pictures. The chief spirit of this organization was a Jesuit priest, Fr Richard Devane, who had been known to confiscate British publications from recalcitrant newsagents in his native Limerick and, with his helpers, set fire to them publicly. He was ably assisted by the Irish Christian Brothers, a religious order dedicated to the education of Catholic boys, for whom they established the periodical *Our Boys* in 1924, partly to offset the spread of British publications for young people. Devane alleged that 'many of the boys' papers are, in effect, recruiting agencies for the British boy scouts which, in turn, are a recruiting agency for the British Army and Navy' (Coogan, 1993: 11).

Both the Vigilance Association and the Christian Brothers turned their attention in 1925 to the new minister for justice, Kevin O'Higgins, firmly believing that a native Irish administration would pay more attention to their entreaties than had the distant government in Westminster (Horgan, 1995: 62). The minister was supportive, but unwilling to move without evidence of widespread public support. His hesitation was cured by a meeting with a deputation of Catholic bishops in January 1926, after which he appointed the Committee on Evil Literature to examine the question (NAI, JUS/7).

Devane, although not a member of the committee, influenced it strongly both by giving voluminous evidence of wrongdoing by imported journals, and by writing copiously on the subject. In one of his effusions, he noted:

> In this question we must remember we are not dealing with the liberty of the Irish press but with the licence of an extern press. Hence I suggest that the proposed legislation be not directed against the home press, but to the outside press. To my mind Irish journalism and the Irish press are as near perfection in this matter as any press can be. (At least 99% of the Irish press. I am not unaware that a few Dublin productions are giving, at times, some cause for anxiety. The Irish provincial press is such as the most exacting could demand in the matter under consideration). (Devane, 1927: 16)

The *Irish Times* noted 'the smug voice of cant ... in the demand for a moral censorship of the press (which would) merely feed the national vice of self-complacency ... The things which defile Ireland today come not from without, but from within' (*IT*, 19 Feb. 1926). The *Irish Independent*, its mass circulation more at risk commercially from British competition than that of the *Irish Times*, was vociferous in its support for the committee:

> There are stringent regulations to deal with the sale of anything that may prove poisonous to the body; but there is no attempt made by the law to prevent the indiscriminate circulation of imported papers that poison the soul ... The fact that the vilest newspapers are flaunted in the face of the public every Sunday, while no prosecutions ensue, is evidence enough that the present law is powerless, unless, indeed, one assumes there is no desire to enforce it. (*II*, 19 Feb. 1926)

The *Irish Independent* warmly welcomed the committee's report, which was delivered by the end of 1926 and published in early 1927 (and was the foundation for the 1929 Censorship of Publications Act, which was to remain in force without major modification until 1967). The *Independent* said the report's proposals 'no more attempt to interfere with the liberty of the press or of the subject than does the legal code against criminal libel or against bigamy' (*II*, 1 Feb. 1927). The *Irish Times*, on the other hand, went some distance in its attempt to reconcile its liberal Protestant ethic and demonstrably middle-class concerns. Public opinion and 'old fashioned manliness' were more effective censors than legislation, it suggested. Moreover, the proposed Censorship of Publications Board's power could easily be 'perverted to the use of the faddist with an unhealthy mind'. It added:

> We have seen in Ireland attacks upon improper literature entrusted to young people who ought not to know that there is such a thing, and often the crusade has done more to taint innocence than the thing against which it was directed. A healthy child can be loosed among the classics without danger; but the little prig who is taught to peep for matter to denounce in every print can hardly fail to be infected. (*II*, 1 Feb. 1927)

When the Dáil debated the report, the moral majority was evident on all sides. The minister for justice, Mr James FitzGerald-Kenney (his predecessor, O'Higgins, had been murdered by the IRA in the interim), told parliament:

> We will not allow, as far as it lies with us to prevent it, the free discussion of this question which entails on one side of it advocacy. We have made up our minds that it is wrong. That conclusion is for us unalterable. (DD, 18 Oct. 1928)

The Censorship of Publications Board, which was set up as a result of the committee's report, had a major effect on the development of Irish literature over the next three decades. The main effect of the legislation insofar as the newspapers were concerned, however, was to make the importation of any material containing information about contraception a criminal offence. For many years thereafter, this legislation was operated on a class-specific basis. It was used, for example, to proscribe British newspapers like the *Sunday Chronicle* and *Reynolds News*, which carried small advertisements for family planning requisites. Another paper to fall under the ban was the *Daily Worker*, whose political sentiments were unwelcome but not illegal; its small advertisements for birth-control materials were a convenient excuse for banning both message and messenger. On the other hand, publications like the *New Statesman* and the *Spectator*, which carried identical advertisements, never attracted official attention in the same way, presumably because their middle-class and better-educated readers were assumed to be impervious to such temptations.

As far as broadcasting was concerned, the establishment of the BBC in 1922 provoked interest, and an early attempt at emulation, in the infant Free State. In much the same way that in the 1960s the national airline became a badge of independence and self-sufficiency for a number of former colonies, the creation of a national broadcasting service was to assume a high priority even as the country struggled to extricate itself from the huge cost of the Civil War, which ended in May 1923.

As it happened, the Treaty that had established the Free State, and that had led directly to that civil war, also had a bearing on the establishment of a radio service. This was because the British negotiators, conscious as ever of Ireland's place in the scheme of imperial defence, had included a clause (not published at the time) restricting the right of the new government to establish radio stations with a capacity for broadcasting outside the national territory. Any such stations had to have prior British agreement. This was to become relevant later, when the power of broadcasting apparatus was increasing rapidly, and when a number of commercial organizations sought to use Ireland as a base for their activities.

Many of the early moves into radio were made by entrepreneurs rather than by governments. These entrepreneurs were primarily interested in technology rather than in content: inventors, manufacturers and distributors of wireless receiving apparatus, they conceived of broadcasting as a way of selling their wares, and of broadcasting content primarily as a means to this end. The BBC itself was formed initially by a consortium of commercial organizations that had been operating independent experimental stations, and which were pushed together by the post office. One of these organizations, the Marconi Wireless Telegraph Company, was simultaneously directing its attention to the neighbouring island. Marconi had carried out many of his early experiments in wireless telegraphy in Co. Cork, and in May 1922, when Ireland was in the early stages of its civil war, he applied to the Irish government for permission to set up a station. A similar, almost simultaneous application, was received from the *Daily Express.* This was not proceeded with, and Marconi's was turned down 'owing to the present disturbed conditions in Ireland' (Gorham, 1967: 5). Undeterred, Marconi went ahead with experimental transmissions over a range of some seven miles from Dún Laoghaire, in the southern part of Co. Dublin, to the Dublin Horse Show in August 1923. The government disapproved, and his apparatus was dismantled and sent back to Britain.

As the 'disturbed conditions' improved, rivalry and emulation went hand in hand: the Irish postmaster general (his title, like much else, borrowed from the UK administration) proposed in November 1923 that broadcasting be established as a commercial concern under the effective control of native private enterprise, but doubts about this in the Dáil (the lower house of parliament) prompted, instead, the establishment of a Dáil committee in the following month. The committee, whose work was bedevilled by a scandal involving alleged links between one of its members and a private organization involved in lobbying for the right to set up a radio service, presented its interim report in January 1924 and its final report in March of the same year. In one Dáil debate on the issue, the postmaster general warned that the Dáil risked surrendering Irish broadcasting to 'British music hall dope and British propaganda'; one of his Southern unionist critics replied tartly that 'if we are to have wireless broadcasting established on an exclusively Irish-Ireland basis, the result will be "Danny Boy" four times a week, with variations by way of camouflage' (McLoone, 1991: 3).

The peculiar form the debate on Irish broadcasting took at this time can be explained at least in part by the peculiar nature of the Irish parliament. Those of its members opposed to the Treaty had withdrawn from the assembly just before the beginning of the Civil War. They had not re-entered the arena of constitutional politics, and indeed were not to do so until the Sinn Féin organization split, leading to the creation of Fianna Fáil in 1927 and that party's subsequent rapprochement with the Dáil. In the interim, and in the absence of any political opposition that could unseat the government, individual members had considerably more control over the executive than later became the case. In this case, they were able to prevail, not only over the doubts of the postmaster general, but over the delaying tactics exercised by the department of finance (Farrell, 1991: 13).

This was the context in which the postmaster general found his proposals rejected, and the administration found itself saddled with a broadcasting model, and a broadcasting philosophy, which was in many respects not of its own choosing. The postmaster general was J.J. Walsh, who had, before 1916, been a telegraph operator. He believed he had devised a model that would allow the government to control broadcasting content without having to pay for the staff and technology used to prepare and transmit it, because it would be operated on a concession basis and financed by a licence fee; but it rapidly became apparent that the erstwhile revolutionaries in the Dáil would not be content with anything that fell short of full political control, and were nearly evangelical in their view of the potential for the new service.

Walsh had told the committee of his own belief that 'any Irish station is better than no Irish station at all' (Gorham, 1967: 12), but his enthusiasm notably abated when he encountered the views of the committee. This was because the committee not only recommended strongly that the new station be set up as a state monopoly (with cost implications that Walsh clearly foresaw, and which he felt unable to justify in the light of the financial stringencies being visited on his department in the aftermath of the Civil War), but had a vision of the future content of broadcasting with which he found himself out of sympathy.

The committee's final report was strongly dirigiste in its view of the future. It considered broadcasting:

> to be of incalculable value: it views the use of wireless telephone for entertainment, however desirable, as of vastly less importance than its use as ministering alike to cultural and commercial progress … a lecture, which otherwise would require to be repeated at different centres involving no little inconvenience and multiplied cost can be heard simultaneously in a thousand schoolrooms and in the home of all who desire to learn … In this way the pupils may learn the elementary principles of hygiene, of gardening, of fruit-growing, bee-keeping, poultry-raising and the like direct from men of recognised authority in the subjects. Similarly, expositions of the reasons for and the application of new laws, lessons on the institutions of government and civics generally might be disseminated in an attractive fashion … As compared with its power for promoting such valuable ends the Committee regards the employment of broadcasting for entertainment purposes as quite subsidiary and worthy of consideration, here, rather as a possible source of revenue than as an essential element in the problem. (DD, 28 Mar. 1924)

The asperity with which Walsh greeted this manifesto was undoubtedly related in part to the fact that, on the very day on which the debate took place, the government had had to secure parliamentary approval for a reduction of a shilling in the old-age pension. He warned against accepting the degree of financial responsibility involved, and added, even more darkly, that the only other country in the world which was tempted to choose the path of total state control was Russia. He added:

> The people want amusement through broadcasting; they want nothing else, and they will have nothing else. If you make amusement subsidiary, then you will have no broadcasting, nobody will buy an instrument,

nobody will pay a licence, and the thing will never begin. (DD, 3 Apr.
1924)

In early 1925 Walsh, whose title in the interim had been changed (he was
now minister for posts and telegraphs), brought the outline of his new scheme
to parliament. There would be a main station in Dublin, and a subsidiary one
in Cork (the previous year he had warned parliamentarians that he had
requests from almost every county for its own station). The licence fee, at £1,
would be double the UK equivalent, and although some 2,800 licences had
been purchased even before the new Irish station began broadcasting, the
attempt to persuade the listening public to part with the finance necessary to
support the new station was to be unremitting and frequently unsuccessful.

The technical work on the new service was carried out with considerable
help from the BBC, and John (later Lord) Reith, then managing director and
later (1927) director general of the BBC, actually sat on the first board
charged with appointing a station director (Gorham, 1967: 19). A one-
kilowatt transmission apparatus was set up in a military barracks close to the
centre of the city (the site continued to be used for this purpose until 1980).
The range at which the transmissions could be received depended on the
sophistication of the receiver: a crystal set could pick up the signal up to 20
miles away, a one-valve set up to 50 miles away, and a two-valve set up to 80
miles away. Even a one-valve set, however, cost approximately £2, the best part
of a week's wages for the average worker. An enterprising primary-school
teacher in Achill, some 180 miles from Dublin, proudly claimed that he had
received the test transmissions 'at good loudspeaker strength, and comparable
with Daventry [the BBC 25 kw station] in every respect'. He added:

> It might not be too much to hope that in the near future the education
> department would have receiving sets installed in all the national schools
> in the State, and thus bring into the otherwise colourless lives of many
> poor children the helpful and delightful influences connected with
> listening-in. (*II*, 30 Dec. 1925)

The new station was allocated the call sign 2RN (i.e., 'to Erin') and began
broadcasting at 7:45 p.m. on Friday, 1 January 1926. The opening
announcement was made by a non-governmental figure, Douglas Hyde,
whose evangelism for the Irish language (he was founder of the Gaelic
League) was enhanced by the fact that he came from the country's religious

minority. His speech, the greater part of which was in Irish (after an introduction in English 'for any strangers who may be listening in'), underscored
an important theme in the station's early years, which was to become
something of a battleground in Irish broadcasting later. This was the extent
to which the national broadcasting service should – or could – be used as a
key element in the official policy aimed at reviving Irish as the spoken
language of the majority of Irishmen and Irishwomen, which it had not been
for almost a century.

There was one notable absence from the programmes broadcast on the
opening night: news. The station had still to come to an acceptable financial
arrangement with the news agencies (which were British) or with locally based
newspapers, for the supply of this exotic commodity. The first advertised
news bulletin, on 24 May, may have been a rebroadcast of British news about
the general strike then in progress in the United Kingdom; more frequent
news bulletins featured from the end of June, but a regular daily news bulletin
did not appear until 26 February 1927, when it was broadcast at 1:30 p.m.
(Horgan, 2004).

2RN broadcast initially six days a week, rapidly increasing to seven. The
programme schedule, however, indicated at least a partial victory for Walsh.
There were no broadcast lectures on animal husbandry or personal hygiene;
instead, the station offered a solid diet of music, classical and traditional, the
latter echoing the Victorian drawing-room rather more than the raw,
untutored talent of the folk music that was a vibrant presence in the
countryside.

There was as yet no formal legislative framework to support the new
station financially. The handful of people who ran 2RN were to find, over
and over again, that the dead hand of the department of finance acted to
obstruct or deny the fulfilment of often very modest ambitions. This did not
prevent costs from creeping up, and no matter how much these were
restrained, they lagged increasingly behind licence-fee revenue. With what
sounded like desperation, Walsh told the upper house of parliament, the
Seanad, in July 1926 that only 5,000 licences had been taken out among the
estimated 25,000 people who owned sets, and that the station's revenue needed
to be supplemented by a duty on imported sets and components (Gorham,
1967: 33). The licence fee was halved in August, but there were few prosecutions for evasion, and district justices generally took a lenient view of
offenders, which did not encourage compliance.

A legislative framework for the new system was provided with the passage of the Wireless Telegraphy Act in November 1926, Section 2 of which prophetically defined wireless telegraphy as any system capable of transmitting 'messages, spoken words, music, images, pictures, prints or other communications, sounds, signs or signals by means of radiated electromagnetic waves'. It also put the collection of licence fees on a firmer footing, and by the end of the following year, it was reported that the number of paid-up licences had risen to 15,000:

> It may be that the vigilance of the Civic Guard also must be thanked, in part, on account of the reduction of unlicensed users of 'wireless' receivers. It is likely that considerable numbers of unauthorised persons continue to tap the 'wireless' waves that roll through the Free State, and we hope that these will be roused to a sense of the meanness of their conduct. (*IT*, 5 Dec. 1927)

During 1926, 2RN pioneered the broadcast of live sports events; it is claimed that the transmission of commentary on a hurling match in 1926 was the first ever live broadcast on a field game. News made its delayed appearance on 2RN after the conclusion of satisfactory arrangements with the providers. These included the BBC, which gave permission to excerpt some of its foreign bulletins, and in general the service was merely the recirculation of material already published elsewhere, notably in the *Evening Herald* and the *Evening Mail*. This could be supplemented from unusual sources: one of 2RN's technicians had been a ship's wireless operator, and used to listen to Morse transmissions from ships in the area for interesting snippets of information for rebroadcast (Clarke, 1986: 50). The station's staff was enlarged by one to facilitate the introduction of this service. The employee concerned, known as the 'station announcer', was originally employed on a part-time basis, but by November 1926 he had become full-time, and a planned extension of the broadcast news was undertaken. The minister told the Dáil:

> Broadcasting will develop a line of Press dissemination very much wider than we find at present. Already a Press bulletin is being issued nightly. It includes news from half a dozen foreign capitals and sometimes local doings, and certain local markets. That bulletin occupies seven or eight minutes of our time. It is the intention to extend that margin to perhaps fifteen or twenty minutes later on. (DD, 30 Nov. 1926: 369)

The context in which this announcement was made was a debate on an advertisement for the employment of an additional, part-time reporter. The fact that broadcasting was – and remained until 1961 – an integral part of a government department, was fully reflected in the job description. The unfortunate employee was expected to record the proceedings of Dáil and Seanad, to obtain reports from police and fire brigades of 'important or sensational happenings', to obtain particulars of outstanding events from national cultural organizations, and to prepare a bulletin of foreign news, stock-market reports, provincial news and market prices. In addition to all of this, and on a part-time salary of £4 per week, he would be required:

> to furnish a guarantee that there will be no infringement of copyright, and shall provide a bond for £1,000 to be obtained from an approved insurance society or company indemnifying the Minister for Posts and Telegraphs as representative of the State against any action, claims or expenses caused ... by any infringement or alleged infringement of copyright (DD, 30 Nov. 1926).

Another innovation was the broadcasting of specially arranged symphony concerts. At the first of these, in November 1927, working-class listeners were let in early, at a specially reduced price; middle-class patrons were not admitted until later, and had to pay more. But not all innovations were equally welcome: John Logie Baird, who visited Dublin in January 1927 to lecture on his new ideas about how pictures could be transmitted, proposed to give a talk on 2RN about his invention, but withdrew when he was offered a brief spot at a time when virtually no one would have been listening (Gorham, 1967: 40, 44, 36).

By then, broadcasting was – surprisingly – paying for itself. But there was a potential flaw: the import duties, which were providing the bulk of the revenue, were to dwindle as more and more Irish companies began manufacturing sets within the boundaries of the state. Advertising had begun, but was yielding very little revenue, and there were the seeds of future controversy in arguments about whether more modern kinds of music might be permitted on the Irish airwaves. A critic in one of the four rival periodicals established to tell listeners about Irish and UK broadcast schedules issued a public warning about the new music – jazz – which had crossed the Atlantic at high speed and was sweeping through Europe. Its popularity, he sniffed, was 'not a very sound argument as to why we should not discard – if necessary by decrees – the music of the nigger in favour of products of our own artistic creation and the creation of cultured peoples' (Fanning, 1983: 22).

This was not an atmosphere receptive to new ideas, or technologies. A further brief negotiation between John Logie Baird, one of the inventors of mechanical television, and the new broadcasting station, came to nothing, although the *Irish Times* predicted hopefully in 1929 that experiments with television were just about to take place, and that 'there is a possibility that the transmission of pictures may be included in the nightly broadcasts' (*IT*, 20 Nov. 1929). Few more optimistic forecasts can have been made.

Whatever about content, the popularity of radio was evidenced by the flood of complaints about poor or inadequate reception from outlying areas of the state. The Dáil in 1930 therefore agreed to a proposal by Walsh to establish a new high-power station at Athlone, more or less in the geographical centre of the country. This was at a time when the station's finances were showing signs of strain. Its Cork studios, which had been producing a weekly programme since 1927, were closed down in 1930 at a saving of £1,000 a year, and licences appeared to have reached a plateau, with an increase of less than 1,000 to a total of 26,000 by April 1931.

In 1932, the establishment of a more powerful transmitter in Athlone (usefully contemporaneous with the Eucharistic Congress of that year) made radio broadcasts accessible virtually everywhere in the state. Radio Athlone, or Raidió Áth Luain, as it was more formally described, was officially renamed Radio Éireann in 1938.

Faced with these financial problems, and with the apparent unattractiveness of advertising (it generated only £1,000 out of a total revenue of £45,000 in 1930–1), station managers embarked on a new device to ensure that the station would pay its way without assistance from general taxation. This was all the more necessary in that the new Athlone station was estimated to cost about £58,000, all of which would have to be found out of future revenues.

The commercial innovation spurred by these figures was the sponsored programme, as 'undiluted advertising talks as had previously been allowed were not a desirable form of broadcast advertising' (DD, 23 Apr. 1931). A sponsored programme was one whose production was paid for in its entirety by a commercial undertaking, but in which the announcement of the sponsor's name was limited to a brief section at the beginning and the end of each programme. Preference was given to Irish manufacturers, and importers were not allowed to advertise items that were being sold in competition with those produced by Irish manufacturers. Until radio – and later television – came into its own as an advertising medium, these programmes, often produced with a high degree of professionalism, attracted a substantial and regular audience: they included drama serials, programmes produced by music retailers, and – not least significantly – programmes produced by the Irish Hospitals Sweepstakes, a private company that had been given the power to run a national lottery on a monopoly basis, and which used the increased power of the Athlone station, particularly in the early 1930s, to inform listeners not only in Ireland but in Britain (where the lottery was illegal) of its attractions. In 1935, after advertisements for its lotteries had been banned in UK papers, it even offered to provide the government with a completely free radio station, if it were allowed to broadcast advertisements for its tickets into Britain; despite the enthusiasm for this scheme of the minister for industry and commerce, Sean Lemass, the government turned it down (NAI, Taoiseach: S 7095).

The quiet, almost placid world of mainstream print media, meanwhile, was about to be disturbed. As already noted, de Valera had been actively seeking support and money for his proposal to establish a new daily newspaper that would provide a platform for him and his supporters: all the existing national dailies were opposed to his new Fianna Fáil party to one degree or another, not least because the Civil War was still a fresh memory. The 1927 *ard fheis* (annual conference) of his party passed a resolution urging the Fianna Fáil national executive to explore the possibility of starting a new paper, but this had already been in progress for some time (O'Brien, 2001; Burke, 2005). Irish Press Ltd was formally established in Dublin in September 1928, and 200,000 shares were offered at £1 each to encourage small investors. Battle lines were being drawn, in politics and journalism alike.

In Northern Ireland, where the religious and political divisions in society had been mirrored in the media for many years, the close-knit nature of both communities ensured that the cohesive, bulletin-board functions of both the *Irish News* and the *News Letter* were a bulwark against many of the diversions offered by the British press. A partial exception was provided by the *Belfast Telegraph* (founded as the *Belfast Evening Telegraph* in 1870, dropping the word 'Evening' in 1918), which had by virtue of its monopoly position in the evening newspaper market a readership from both communities.

The *Irish News* had been a proponent of home rule for Ireland within the British empire, and was deeply disappointed by the partition solution embodied in the Treaty. The establishment of the Northern Ireland parliament at Stormont in May 1921 was, it maintained, 'a plan by unscrupulous politicians to assassinate Ireland's nationality'. It quickly came to the conclusion, after the territorial status quo had been effectively ratified by the Boundary Commission in 1925, that Northern Catholics and nation-alists had to work out their own political salvation within the new constitutional arrangements, which angered nationalists (including some bishops) who felt that an all-Ireland solution was still a short-term possi-bility. These two were reunited in 1928 under the nationalist politician Joe Devlin, who had been managing director of the paper since 1923 (Phoenix, 1995: 24–5), but circulation was weak, and was revived only by the appointment of a new editor, Sydney Redwood, in 1929, who cut the price of the paper to a penny, and, by a combination of aggressive marketing and journalistic flair, succeeded in substantially reviving its fortunes and making it an effective voice of mainstream Northern nationalism (Phoenix, 1995: 27–8).

The *Belfast News Letter*, owned by the Henderson family, was no less staunch in its defence of unionism and, before the partition of the island, of the British connection. In November 1919, when the violence of the War of Independence was making itself keenly felt in the North, it allocated the blame unambiguously in religious terms: 'It is the bigotry of [the Catholic] Church, and its constant efforts, open and secret, to increase its power, which have brought a large part of Ireland to the lawlessness which is disgracing it today' (Kennedy, 1988: 40).

In the May 1921 election to the Stormont parliament, the *Belfast Telegraph* printed unionist candidates' speeches in full, accompanied by photographs captioned: 'Vote Only For These Candidates'. When, after the April 1925 election in Northern Ireland, the president of the Irish Executive Council in Dublin, W.T. Cosgrave, sent a telegram of congratulations to Northern nationalists who had been elected, the *News Letter* castigated what it described as the 'direct influence by the head of one state in the internal affairs of another' (Keogh, 1994: 26).

The *Northern Whig*, owned by the Cunningham family, had been a radical paper in the nineteenth century, but the home-rule controversies pushed it firmly into the unionist camp after 1885. If it was solidly unionist, however, it also gave vent from time to time to unusual views. William Armour, who was appointed editor in 1928, was the son of a Presbyterian clergyman and supporter of home rule: his brief tenure of office was characterized by a subliminal sense of disenchantment with unionism.

Broadcasting in Northern Ireland, both in its origins and in its modalities, reflected the political and social anomalies of the region. As Rex Cathcart points out in his definitive history of the BBC in Northern Ireland up to 1984, there was at first no compelling reason why Northern Ireland should have had a separate BBC presence: no other region of the United Kingdom had one, and Northern Ireland's population was only 1.5 million, much smaller than that of Scotland or Wales (1984: 5). Local enthusiasm, however, carried the day, and a decision to plan for the new station in Belfast was taken in December 1923 and announced publicly in February 1924. The opening broadcast was transmitted on 15 September 1924, but the transmitter was of comparatively low power, with an effective range of not much more than 30 miles. Until the power was enhanced, there were frequent complaints from would-be listeners outside this radius, notably in Derry. Although there was a stipulation by the unionist government of Northern Ireland that it should retain the right to terminate transmissions from the Belfast station at any time if it thought this necessary in the public interest, initial reaction to the programming from within both communities was broadly welcoming.

The opening of 2RN in Dublin in January 1926 did not seem to raise the temperature perceptibly. Indeed the Belfast and Dublin stations co-operated on a number of musical programmes, which were generally the subject of favourable comment. The honeymoon, however – if this is what it was – was to come to an end for a number of reasons, some of them internal to Northern Ireland itself, others emanating from south of the border.

Despite the early enthusiasm of the *Irish News* for the new Belfast station, the Catholic church in Northern Ireland declined to participate in the workings of the committee set up by the BBC to advise on religious broadcasts there, and maintained this stance until 1944. Similar tensions began to obtrude in the form of controversies about the arrangements made by BBC NI for St Patrick's Day each year. Neither community could ever be entirely satisfied. Unionists, in particular, were more particularly exercised by the fact that these particular programmes were being transmitted for broadcast throughout the United Kingdom, and were therefore labelling them culturally in a manner of which they disapproved at some visceral level (Cathcart, 1984: 44–5).

Plans to increase the level of co-operation in programming between Belfast and Dublin were an early casualty of this new atmosphere, and growing inter-communal tension in the mid-1930s sharpened a growing awareness of the political significance of broadcasting, especially where news was concerned.

Although party political broadcasts were inaugurated in Britain for the 1933 general election, none were allowed in Northern Ireland before the war. There were serious disturbances in Belfast in 1935, involving several deaths, and the BBC's practice of reporting these without analysis or context provoked one unionist politician to argue in the *Northern Whig* the following January that direct control of BBC NI by the Stormont government would make possible 'counter-propaganda to the republican stuff which comes over nightly from Athlone' (31 Jan. 1936). To its credit, the *Whig* rejected this analysis (1 Feb. 1936).

Charles Siepmann, the BBC's director of regional relations, was sent to Belfast from London in 1936, as preparations were being made for the establishment of a fuller regional service in Northern Ireland in that year. He wrote a much-quoted report in which he expressed his view that broadcasting in Northern Ireland was necessitated by political considerations, and was not justified by the availability of indigenous programme resources. He added:

> The bitterness of religious antagonism between Protestants and Catholics invades the life of the community at every point and for our purposes conditions almost everything we do ... In the official life of Northern Ireland the Roman Catholic by virtue of his religious faith is at a discount ... Politics and public administration conditioned by religious faction have about them an Alice-in-Wonderland-like unreality. (Cathcart, 1984: 3)

A year later, the regional director in Belfast, George Marshall, was to set out his own position in a quite uncompromising form when, objecting to a programme entitled 'The Irish' that had been produced by the BBC's Northern Region, he observed: 'there is no such thing today as an Irishman. One is either a citizen of the Irish Free State or a citizen of the United Kingdom of Great Britain and Northern Ireland. Irishmen as such ceased to exist after the partition' (Cathcart, 1984: 6).

The stage was set: the play would have a long run.

3 Affairs of state, 1932–47

The years between 1932 and 1948 were years of uninterrupted rule by Fianna Fáil. This was not paralleled by placidity or quietism in the media. The establishment of the *Irish Press* in November 1931 by the leader of Fianna Fáil, Eamon de Valera, served notice on the existing media establishment that things would never be the same again. The outbreak of war in 1939 created a whole new series of tensions between media generally and the government, and, in the post-war period, radio began to come into its own as a public service with a much broader remit than hitherto.

The establishment of the *Irish Press* was as seismic an event, in its own way, as the arrival of 2RN. It was unique in twentieth-century Ireland as the only national newspaper with an overt relationship to a national political party. In this it mirrored similar newspapers in a number of Continental countries, where the tradition of a party press, and of a press linked to trade unions, was well established. The similarities, however, ended there: the *Irish Press* was in many respects the unique creation of its progenitor, Eamon de Valera.

The principal objective of the paper was to focus and co-ordinate electoral support for Fianna Fáil. But it also had to be a commercial enterprise: no party could support a dependent national newspaper for long. Achieving these objectives meant acquiring journalistic and commercial credibility within a structure that would also guarantee its political loyalty.

This balancing act found formal expression in the articles of association of the company established to publish the paper. The board of directors comprised de Valera and seven others – prominent businessmen, chosen not only because their companies operated across Ireland, but also on the basis of religion (one director was a prominent member of the Church of Ireland). These choices were intended to reassure the business community – whose advertisements would be needed to support the new publication – and to convince Irish Protestants that republican journalism would not be in any way sectarian.

The political loyalty of the paper to Fianna Fáil was underwritten by another, even more important, structural element. This was the designation of de Valera as controlling director, with the power to assume any or all of the functions associated with the management and control of the newspaper. This was a sophisticated formula. The paper would not be controlled by Fianna Fáil but, as leader of the party, de Valera could act whenever necessary to ensure that it did not operate against the party's interests. Setting the paper at arm's length from the party had two distinct advantages: it created a veneer of political independence while protecting the paper and its journalists from the kind of party interference that would have probably guaranteed its commercial ruin.

Its first issue, on 5 November 1931, carried an editorial that declared that the *Irish Press* was 'not the organ of an individual, a group, or a party', but would be 'a national organ in all that term conveys'. It added:

> We have given ourselves the motto: Truth in the News. We shall be faithful to it. Even where the news exposes a weakness of our own, or a shortcoming in the policies we approve, or a criticism of the individuals with whom we are associated, we shall publish it if its inherent news values so demand … Until today the Irish people have had no daily paper in which Irish interests were made predominant … Henceforth other nations will have a means of knowing that Irish opinion is not merely an indistinct echo of the opinions of a section of the British press.

Its first editor, Frank Gallagher, had a pedigree in nationalist journalism that included work on the Sinn Féin *Bulletin* during the War of Independence. His instructions to reporters and sub-editors as the paper prepared for its launch date made its ideological direction quite clear (NLI, MS 18336). Its hostility to the prevailing administration was embodied in a directive that it was not necessary to report every word of praise spoken by judges to policemen – courts and police alike were actively involved in the suppression of the remnants of the defeated Civil War forces and, frequently, the harassment of their sympathizers. At the same time, they were reminded not to make the *Irish Press* merely a Dublin paper, and to emphasize aspects of news stories featuring women and children. Sub-editors were warned that 'propagandist attacks on Russia and other countries should not be served up as news' – an indirect admonition to be wary of the value systems embodied in the British and American news agencies, virtually the only available sources of international news.

From the start, the new paper made it clear that its primary target readership was composed of the two groups providing Fianna Fáil with most of its electoral support: the urban working class and small farmers. It developed an immediate expertise in reporting sport, particularly Gaelic football and hurling. It was virtually alone in publishing much writing in the Irish language (although this component was to shrink in the years ahead), and provided an outlet for a generation of gifted and eager young journalists and writers effectively excluded from media employment for years by virtue of their political views. It also broke new ground in that it was the first daily morning paper to put news on its front page.

Public reaction to the new paper was initially enthusiastic: its print run of 200,000 for the first edition sold out, and it eventually settled down to around 100,000. The Cumann na nGael administration did not disguise its hostility, banning *Irish Press* reporters for a time from party meetings and declining to supply it with information in the normal way. Existing newspapers responded in different ways to being described as 'a section of the British press'. The *Irish Times* gave it a patrician welcome, even reprinting an extract from its first editorial. The *Irish Independent* carried on as if the *Irish Press* did not exist, and edited out references to the paper from its reports of speeches by Fianna Fáil politicians.

The Murphy family, the major shareholders in the Independent group of papers, also used their commercial muscle to veto a request by the *Irish Press* to be accommodated on the special newspaper trains taking early editions of the morning newspapers to non-metropolitan destinations. This was facilitated by the interlocking directorships of the chairman of the Independent board, Dr Lombard Murphy, who was also a director of the privately owned Great Southern Railway Company. The *Irish Press* eventually succeeded in appealing this policy to the railway tribunal arbitration body in 1932 (*IP*, 22 Apr. 1932), but the need to set up a completely independent distribution network in the interim meant heavy initial costs.

Although the head of the Irish government, W.T. Cosgrave, brushed aside suggestions from his more extreme followers that the new paper should be banned – it would, he thought, wither away naturally (Oram, 1983: 175) – it rapidly became an irritant that could not be ignored. This was particularly because Gallagher, refreshing a technique he had employed to considerable effect during the War of Independence, began to publish articles containing cogent evidence that members of the Garda Siochána had treated political opponents of the government with brutality.

This was a new element in an already highly volatile political situation. The economy, devastated by the Civil War, was in poor shape. Fianna Fáil, having abandoned its abstentionist policy, was now participating in constitutional politics, and aiming at replacing the government at the next general election.

Its commitment to the democratic process was, however, frequently questioned by the government, which reacted strongly to published questioning of the state's legitimacy or hints at the validity of extra-parliamentary action. This led directly to the passage in October 1931 of the Constitution (Amendment No. 17) Act, establishing a military tribunal with wide powers to suppress publications. Although the tribunal's proscriptions included the republican newspaper *An Phoblacht*, its activities were directed to a greater extent against left-wing minority publications, such as the *Irish Worker*, edited by the trade-union firebrand – and communist – Jim Larkin, and the *Worker's Voice*, whose banned issue of 31 October 1931 declared that 'the form of the State is a secondary matter – the issue is which class rules' (NAI, Taoiseach: S 4669).

Having implied that the police were acting beyond their legal authority, Gallagher's *Irish Press* articles immediately came under official notice and, in December 1931, he was arraigned before the tribunal on a charge of publishing seditious libel. He defended himself vigorously, but the 1932 election intervened before the case had been concluded. The election saw de Valera form a minority government while the newspaper of which he was controlling director was still before the tribunal. Gallagher was eventually convicted, and fined £50. The fact that his trial was in progress during the election campaign helped to enhance support for Fianna Fáil at the polls, but the role played by his newspaper was undoubtedly much more significant. The *Irish Independent* commented, with all the grace of a bad loser:

> Yesterday the Fianna Fáil party's organ indulged in quite an orgy of abuse of other newspapers for what was termed 'misrepresentation' in regard to the elections ... On the same day that Fianna Fáil worked itself into a paroxysm of vilification, that journal and its editor were convicted of publishing a seditious libel with intent to vilify the Irish government and the National Police Force. (*II*, 18 Feb. 1932)

Political advertising in all the newspapers was a notable feature of the election. Despite having the *Irish Press* at its back, Fianna Fáil advertised in both the *Irish Times* and *Irish Independent*, countering claims made by its political opponents in advertisements in the same papers. Neither of the two government-supporting papers allowed their editorial opinions to be swayed by the receipt of these funds: in its pre-election editorial, the *Irish Times* warned the electorate that Fianna Fáil land policy was Marxist. After the election, the

Irish Times' opposition to the new administration, unlike that of the *Irish Independent*, was qualified by a willingness to give de Valera a chance to demonstrate he was not the socialist ogre they had earlier presented to their readers. In effect, a double conversion was taking place: de Valera was coming to terms with the compromises necessary in the administration of a country of whose constitutional arrangements he totally disapproved; and the *Irish Times* was coming to terms with a political party whose antecedents it regarded as radically subversive but which was now in government.

Although to a considerable extent the work of government kept de Valera away from the business of running the paper this did not mean that controversy about the paper's political role abated. Indeed, it was centrally involved in three key sets of circumstances: the crisis involving the role of the governor general (the legal representative of the crown in Ireland under the Free State constitution); the discovery by de Valera of a legislative device by which he could refinance the paper; and a challenge by the paper's political correspondent to the still extant military tribunal.

The peculiar relationship between the head of government and his newspaper was publicly debated when the governor general, the nationalist politician Eoin McNeill, took umbrage at an *Irish Press* report about a diplomatic incident involving him, and accused de Valera of having inspired the report. De Valera replied that the *Irish Press:*

> gathers its news as other newspapers do, and publishes what it gathers at the discretion of its editors. Any particular item of news would, it is true, be suppressed were I to issue an express rider to that effect, but ever since the paper was founded I have carefully refrained from giving any such orders. (*IP,* 12 July 1932)

It is important to note what was here being asserted, and what was not being denied. De Valera's interpretation of censorship was a narrow one, and negative. It did not include positive action by him to shape the content of the paper in a way that did not involve direct suppression:

> As a political reporter in the old, barn-storming days of the *Press,* my father was close to Dev. He used to tell me stories about Dev phoning the newsroom every night to know what was coming out in the paper the next day. He often had no hesitation in changing copy, in changing a whole front page. (Kilfeather, 1997: 22)

De Valera's biographers noted that he frequently delegated work, but 'to whatever he took personal charge of he gave the closest supervision'. This included, on the night of the first edition, supervising the despatch of the early editions to the rural areas. They also note his comment that if it ever came to a choice between leading the party and running the paper, he believed that he would choose the latter (Longford and O'Neill, 1970: 270–1).

The crisis involving the governor general reached a climax during the Eucharistic Congress, a major religious festival. De Valera had ensured that the governor general – who, according to protocol, would have taken precedence over the head of government – was not invited to any of the functions associated with this event. After protesting ineffectually, McNeill released the correspondence to the press on the evening of 10 July 1932. In the early hours of the following morning, newspapers were informed that the government had declared the letters to be confidential state documents, which could not be published. Police intercepted British papers at the mail boat, and Northern Ireland papers at border crossings. By the afternoon of 11 July, however, the government realised that as the letters had already 'appeared in foreign newspapers, the Executive Council has decided to authorise publication in the Irish Free State of the entire correspondence' (*IP*, 12 July 1932).

The next day's *Irish Press* drew the obvious moral from the story – obvious to its controlling director, at any rate – charging that McNeill was 'wholly unsuited to the position he occupies' (*IP*, 12 July 1932). The *Irish Times* chose to interpret it as an attempt to muzzle the governor general, rather than as an exercise of press censorship. The *Irish Independent* described the government as resembling 'a lot of spiteful schoolgirls', and suggested that its attempt at censorship had been 'extraordinary and ineffective'. It was left to Northern Irish and British papers to beat the drum loudest in defence of press freedom: the *Northern Whig* regaled its Protestant readers as they gathered for that day's Orange Order marches with a description of de Valera's action as 'an almost incredible display of peevish narrowness', and the *Mid-Ulster Mail* pictured him as 'donning the mantle of Lenin' (11 July 1932). It was left to the London *Star* to point out the moral:

> Even the sternest rebel could catch a hint of the world's laughter that was going to greet the fact that one of the letters which Mr de Valera was forbidding the press to publish was one written by himself in defence of the liberty of the press ... when that press happens to be his own. (12 July 1932)

An equally anomalous position developed at the end of 1932, when the military tribunal, its activities now directed at the paramilitary supporters of the ousted government rather than republicans, heard evidence from Joe Dennigan, the *Irish Press* political correspondent. Having refused to name an anonymous informant quoted in an article on government policy towards illegal organizations, Dennigan was found in contempt of the tribunal and sentenced to a month's imprisonment – the first journalist in the modern Irish state imprisoned for refusing to reveal a source. This evoked a rare display of unanimity among his fellow journalists on all the main papers, who unsuccessfully petitioned de Valera, in his capacity as head of government, to exercise the prerogative of clemency in respect of his employee. De Valera declined, but paid a cannily timed visit to Mountjoy Gaol, after which, Dennigan subsequently reported, the prison diet improved considerably. On Dennigan's release after three weeks imprisonment (with remission for good behaviour), in January 1933, Dublin journalists organized a dinner of welcome for him, and even persuaded the notoriously parsimonious *Irish Press* management to pay for it.

De Valera was not slow to turn his new power and position to advantage in other ways, and now moved to put the final piece in place in a complicated jigsaw begun several years earlier. In 1919–21, Irish emigrants in the United States had subscribed funds for bonds to be redeemed by an independent Irish government. A large amount of money was collected and lodged in a US bank, but, after the Civil War, a dispute arose between de Valera (who had been involved in the fundraising) and the Free State government as to who owned the money. In 1927, the US supreme court decided that it belonged to neither, and ordered that the money be returned to the original subscribers.

In 1930, de Valera wrote to those subscribers, asking them to transfer their right to be reimbursed to him, so that he could launch a national Irish newspaper. Many did so. Now in government, he moved financial legislation to repay all the American lenders – those who had transferred their rights to him as well as those who had not – with a premium, effectively returning $1.25 for every dollar they had lent. The funds for this operation came from the Irish exchequer, which his party now controlled. Despite parliamentary accusations that de Valera was 'looting the public purse for a party organ' (DD, 5 July 1933), the measure was passed: £1.5 million was paid to the bondholders, £100,000 of which found its way directly back to the *Irish Press* on the basis of the transfers signed three years earlier.

De Valera also worked swiftly to improve the terms of trade for all Irish newspapers in a way that responded simultaneously to nationalist and to moral arguments, and reflected his government's strongly protectionist economic policy. In 1927, Fr Richard Devane had argued in the journal *Studies* for a tariff on imported newspapers and magazines (Devane, 1927). Almost as soon as he assumed power, de Valera moved to impose tariffs on some 200 imported newspapers and periodicals. The combined circulation of English daily newspapers in Ireland at this time would have been about 170,000, as against a circulation for Irish dailies of approximately 280,000. What the moralists seem to have been particularly concerned about were the British Sunday newspapers, some 350,000 of which were sold in Ireland every week. De Valera's measures had a marked effect: the value of imported newspapers fell from £216,000 in 1932 to £99,000 in 1934 (O'Donnell, 1945a: 392). The tariffs were maintained until July 1971, when they were eliminated in preparation for membership of the Common Market.

If de Valera's tariffs were indirect aid to the domestic media, more direct aid was also contemplated. In 1934 the cabinet authorized the establishment of an interdepartmental committee to examine the development of a national periodical press, and whether it might be possible 'to restrict the importation of newspapers and periodicals'. This initiative came from the minister for industry and commerce, Sean Lemass, based 'mainly on moral and cultural grounds, though the possible economic advantages, e.g. increased employment, have not been lost sight of' (NAI, Taoiseach: S 2919A). This initiative withered on the vine, and no direct subsidies of any kind were made available to any Irish publication until 1948.

Northern Ireland's unionist papers maintained a high level of watchfulness. The *Belfast News Letter* greeted de Valera's election with predictable dourness: 'It will be well if the people ringing their bells today are not found wringing their hands tomorrow' (Keogh, 1994: 64). The *Northern Whig* had accurately discerned (4 Oct. 1932), in the 1932 imbroglio between de Valera and MacNeill and in the subsequent forced departure of MacNeill, the seeds of the dissolution of the constitutional link between Ireland and Britain. Both it and the *News Letter* (4 Apr. 1935) continually drew their readers' attention to de Valera's gradual dismantling of the British connection. These developments – and the harsh economic climate, which was responsible for widespread unemployment and hunger – provoked in turn more dramatic events, as when, during this period, a large number of mourners at a Protestant funeral were incited to make an attack on the offices of the *Irish News*, and were prevented from carrying out their plans only by the sudden appearance of an armoured car sent by the police (Oram, 1983: 183).

De Valera ensured that a close watch was kept on republican newsletters and magazines, most of which were under permanent scrutiny by the department of justice, and some of which were intermittently banned (NAI, Justice: S Files, 1932–9). The IRA was proscribed in 1936, a development warmly welcomed by the *Irish Times* (Gray, 1991: 65). Indeed, that paper displayed an increasing warmth towards de Valera. Though it advised voters to support the Cumann na nGael party in the 1937 election, a year later it was to be found describing the Fianna's Fáil victory in the snap election of 1938 as 'eminently satisfactory'. 'We are glad', it commented, 'that he [de Valera] has been returned to power' (*IT*, 21 June 1938; Lee, 1989: 215).

Given the expressed support of all the major political parties, but especially of Fianna Fáil, for the Irish language, it is surprising that the creation of the Free State, and de Valera's accession to power, did not see a flowering of national or periodical journalism in the Irish language, then spoken habitually by no more than 10 per cent of the population. De Valera saw to it that his *Irish Press* frequently published material in Irish, but not even his enthusiasm could maintain the output at the level at which it was inaugurated, with a promise of half a page a day. One Irish-language periodical, *An tUltach* ('*The Ulsterman*'), had been in existence since 1924, but nationalism was always strongest in that province for political and historical reasons. The task of reviving the language as the main medium of communication for the majority of the population – an ideological objective all too frequently advanced by people with little knowledge of psychology, linguistics or education – was entrusted chiefly to the schools.

Given that support for the Irish language was strongest among the more conservative sections of the population, it is notable that the most vibrant expression of writing in Irish came from the radical left. *An tÉireannach* ('*The Irishman*'), published weekly from 1935–7, was unique in many ways. It was politically inspired by a small group of left-wing republicans discontented with the sterile opposition of the dwindling paramilitary IRA, and increasingly interested in social issues. In 1935, they formed an organization known as the Republican Congress, a determined but ultimately unsuccessful attempt to win over militant republicanism to the cause of socialism. *An tÉireannach*, their house organ to all intents and purposes, published lively articles excoriating Irish conservative politicians and – unusually for any Irish publication at the time – drawing attention to the increasing menace of Continental fascism (Ó Cíosáin, 1993: 100–4).

In the mainstream press, subtle but significant changes of personnel were taking place. In 1934, Frank Gallagher resigned as editor of the *Irish Press*, frustrated by the penny-pinching attitude of his directors; de Valera found another position for him as head of the government information bureau. A series of unsatisfactory successors ended in 1937 with the appointment of the paper's London editor, Bill Sweetman, a tough journalist who was prepared to stand up to anyone, including, on occasion, de Valera himself. At the *Irish Times*, the editor was John Healy, who had filled this position since before the establishment of the Free State and had engineered a relatively positive approach in his former proto-unionist paper to the new constitutional arrangement, as well as a more lukewarm one to de Valera's accession to power. He died in 1934, and he was succeeded by R.M. Smyllie, an ebullient journalist who had joined the paper in 1919. Smyllie had been working on a provincial paper in Co. Sligo and was on holidays in Germany when he found himself caught up in the First World War, and was interned by the Germans for its duration. Emerging unscathed at the end of hostilities, he was invited to report the Versailles peace conference for the *Irish Times*, and subsequently rose rapidly through the paper's ranks. He was a long-term observer of Germany, about which he wrote regularly and perceptively. On 3 July 1934, for instance, he remarked that 'For the moment, Herr Hitler and Herr Goering control the situation, but the measure of their authority is the quickness of the Black Shirts' trigger fingers' (Schulz, 1999: 119).

At the *Independent*, the new editor from 1935 was Frank Geary, who had joined this paper in 1922. He devoted his considerable journalistic talents to consolidating his newspaper's grip on the growing Irish Catholic middle classes, blending straightforward journalism with an unswerving loyalty to the Catholic church and its moral values.

Editorial changes occurred discreetly. Smyllie's appointment went unmentioned in the pages of his own newspaper (Gray, 1991: 57). He inherited, along with his editorial responsibilities, a necessary sinecure. As Healy had been, he was to be the Irish correspondent of the London *Times*, and in time also of *The Observer*. Although there was no commercial or organizational connection between these papers, this unofficial linkage had been established for many years. Nor was it merely *pietas*: Irish journalists were extremely badly paid and, given that Dublin in some respects shared an information ecology with London, senior editorial executives on Irish newspapers (with the exception of the *Irish Press*) tended to draw a significant part of their remuneration from these external activities. In many cases, it cost them little effort. Occupying key positions in their own news organizations, they simply filtered the Irish news that was passing across their desks and recycled it, when suitable, for the UK editors. This financial nexus was to prove a major factor in the demise, many years later, of the Irish News Agency.

The one international issue of the later 1930s that excited general attention in Ireland was the Spanish Civil War. Here the newspapers were divided on unexpected lines. In the North, the *Belfast Telegraph* saw the war as overwhelmingly one with political rather than religious connotations, and supported the British government's policy of neutrality. The *Irish News* also supported neutrality (which, in essence, was de Valera's position), but from a different perspective, arguing on 30 September 1936 that 'Irishmen can gain for their own country a high reputation if they concentrate on their own and the country's needs and leave cloak-and-sword romance to the novelists' (Bell, 1969: 152).

Given its Catholic readership, the *Irish Independent* adopted a strongly pro-Franco line, making much of reports of atrocities committed against priests and nuns by the opposing republican forces and sending a journalist to cover the war from the Franco side. 'All who stand for the ancient faith and the traditions of Spain', it announced on 22 July 1936, 'are behind the present revolt against the Marxist regime in Madrid'. Local newspapers tended to follow the *Independent*'s line. Small socialist papers like *The Worker* and the Communist Party's *Irish Democrat* urged intervention on the side of the Spanish government, but without conspicuous success. The *Irish Times*, equally predictably in the light of Smyllie's strongly anti-fascist views, was much more measured, publishing 'some of the most factual, balanced editorial analyses to be found in Europe' (Bell, 1969: 140), and treated the legal Spanish government as deserving of support until its final collapse. It sent a reporter, Lionel Fleming, to report from the republican side. Fleming's even-handed despatches, which detailed the anti-religious campaign by the government but also described its actions as a legitimate struggle against fascism, generated a threat of a religiously inspired boycott by several prominent advertisers (Fleming, 1965: 170). The threat materialized, to the point where all Catholic-owned schools withdrew their advertisements from the paper, and Smyllie was forced to withdraw Fleming (Cooney, 1999: 91).

The attitude of the *Irish Press* was intriguing. It did not hide its concern for the victims of atrocities, or for the welfare of the Catholic church, but reported the conflict in a considerably more even-handed manner than the *Irish Independent,* which was remarkable given its conservative and Catholic readership. Behind this was de Valera's political stance. Whatever his personal and religious sympathies he was also, and in this case primarily, the head of a government that recognized the republicans as the democratically elected and thus legitimate government of Spain. This prompted support for a policy of neutrality, which he followed throughout the war and which found a ready echo in the pages of his own newspaper. This was no mere concern for diplomatic niceties, but a conscious decision that his newspaper would not lightly set aside the democratic decisions of the Spanish people. It also came easier to *Irish Press* journalists, many of whom came from a radical Irish republican tradition, whose political leaders had been excom-municated by the Irish Catholic bishops during the 1922–3 Civil War (Bell, 1969: 143).

This was not the only lapse from the standards that religious monitors, hoping for better things in a newly independent Irish state, had been expecting:

> From the secular point of view our newspapers are often little better that the English papers on which they are modelled. They have not the crude sensationalism of the English yellow Press, but great prominence is given to sensational features – murders, accidents, disasters and so forth. Like the English papers they devote a very large proportion of their space to sport. A Dublin daily paper intended for all Ireland will devote half a page to a dog show or a football match, and less than a quarter of a column to some meeting at which the leaders of intellectual life have spoken on a vital topic. (Browne, 1937: 262)

As rumours of wars darkened the European horizon, de Valera pushed forward doggedly with his policy of disengagement from Britain. Amid the crisis surrounding the abdication of Edward VIII, he discovered a legislative loophole enabling him to sever Ireland's connection with the crown (though not with the commonwealth). He then set about drafting a new constitution, which would be debated in parliament before being submitted to the people in a referendum.

The section of the draft constitution dealing with the liberty of the press provoked a lively debate. It guaranteed 'the right of the citizens to express freely their convictions and opinions', but added:

The education of public opinion being, however, a matter of such grave public import to the common good, the State shall endeavour to ensure that the organs of public opinion, such as the radio, the press, the cinema, while preserving their rightful liberty of expression, including criticism of Government policy, shall not be used to undermine public order or morality or the authority of the State. (Article 40.6.i)

This aroused fierce opposition from members of parliament. Trinity College TD Dr Robert Rowlette argued that it would preserve the right of a free press only 'as long as it does not differ with the opinion of the government' (DD, 2 June 1937). John A. Costello, a prominent opposition member and senior counsel, moved to delete the entire subsection as an unwarranted interference with liberty of expression.

De Valera portrayed himself, cannily, as a moderate between two extremes. The press, he reminded Costello, could never be allowed total freedom:

I say that the right of the citizens to express freely their convictions and opinions cannot, in fact, be permitted in any state. Are we going to have anarchical principles, for example, generally propagated here? I say no ... You should not give the proponents of what is wrong and unnatural the same liberty as would be accorded to the proponents of what is right. (DD, 2 June 1937)

Costello was forced into public agreement, even disclosing his personal view that 'the propagation of Communistic doctrine would be against public order and morality' (DD, 2 June 1937). A politically independent parliamentarian, Frank McDermott, defended the draft constitution as a comparatively liberal document – which, in the social and religious context of 1930s Ireland, it undoubtedly was – and attacked the illiberal attitudes on freedom of expression found in the *Irish Independent* and the *Irish Catholic*, and in neo-fascist organizations such as the Christian Front. The latter body, he noted, had been passing resolutions 'calling on the government to make it a crime for a man to express Communistic opinions in his own home' (DD, 2 June 1937). The leader of the small Labour Party, William Norton, expressed the view that the proposed article had 'a real Nazi ring about it' (*IT*, 4 June 1937). The fact that Norton was the only party leader without a national paper to support him shaped de Valera's reply. The purpose of the article, said de Valera, was to provide safeguards against the emergence of 'wealthy people making still more wealth through the Press and making that wealth in a manner dangerous to the public good' (DD, 2 June 1937).

After a debate that evoked considerably more heat than light, the new section was passed. It received an immediate imprimatur from the prominent Catholic theologian Dr Cornelius Lucey (later bishop of Cork), who suggested that the constitution's outlawing of blasphemy might have been improved by the further proscription of 'anti-religious' matter, but that it was otherwise as close to perfect as was possible. In a contribution to the debate that blended a surprisingly contemporary suggestion that public figures should have *less* protection from the laws of libel than ordinary citizens (in order that wrongdoing could more readily be exposed) with a certain steeliness about the dangers of theological error, he observed:

> The right to publish what we may call 'the truth in the news' and 'the views in the news' has one essential limitation. It is subject to the general moral law. Now, the moral law forbids us to say or do anything, either wrong in itself or calculated to lead others into wrongdoing. And the reporter and editor are bound by this law just as strictly as the private individual is. Hence it is not permissible for them to write up what is immoral or uncharitable or otherwise wrong, even on the plea that it is true. For instance, they may not report all the lurid details of divorce proceedings, or set themselves out to promote class-antagonism, or give publicity to the secret failings of this or that private citizen … There is good reason in this country why the State should tolerate all religions and allow them freedom of expression, but there is no reason why it should tolerate unbelief or, at any rate, open propaganda on behalf of unbelief. (Lucey, 1937: 588–90)

Lucey warned about the threats to press freedom in Italy and in Germany, but the dangers he perceived there affected more than the press – the outbreak of the Second World War was imminent. This would present the government and media with a major problem: censorship.

The threat of hostilities did not of itself prompt the government to legislate for censorship. Both the military and the bureaucracy had been nibbling away at the problem intermittently since the end of the Civil War. In 1925, the army authorities prepared a detailed memorandum outlining a scheme for news censorship in the event of an outbreak of hostilities (O'Halpin, 1999: 209), although it is not clear who they thought the protagonists might be. In

December 1930 the cabinet agreed to establish a committee to create machinery for the operation of wartime censorship, with representatives from the departments of defence, external affairs, justice and posts and telegraphs (NAI, Taoiseach: S 8202). The absence of actual conflict induced a certain lethargy, and in October 1935 the executive council was informed that the committee had ceased to function, and was asked to revive it. The committee, now including the attorney general, finally produced a report in September 1938 (NAI, Taoiseach: S 10829). This, together with the Emergency Powers Act of 1939 – rather than the provisions of the 1937 constitution, as de Valera's political opponents had feared – formed the basis for the detailed policy introduced at the outbreak of the war. Both the Emergency Powers Act and the 1939 Offences Against the State Act, which proscribed the IRA (some elements of which were pro-German), gave the state an extraordinary armoury of powers under which many civil liberties, including the freedom of the press, could be subject to more or less arbitrary limitation.

Already in 1938 de Valera had put censorship third in a list of political priorities (after food and external trade, and before counter-espionage and military measures) designed to safeguard Ireland's neutrality. The censorship he now instituted under ex-minister Joseph Connolly, who operated in close conjunction with military intelligence, was based on a carefully worked-out, if occasionally inconsistent, set of imperatives, and was implemented with increasing severity until the very end of the war.

The censorship system was primarily aimed at suppressing statements that might encourage belligerents to conclude that Ireland had abandoned its neutrality and could be classed as an enemy. A secondary objective was to prevent the publication of information of strategic use to the combatants. The censorship, however, operated in tandem with two other policies, at least one of which seriously compromised Irish neutrality, and the other of which made the domestic censorship concerns appear exaggerated.

The first of these policies was that of close co-operation with the Allies, in secret, on a range of subjects that were of importance to the war effort (O'Halpin, 1999: 225–31). These included the exchange of vital intelligence information, meteorological information and coastguard reports. The second was that censorship restrictions did not apply to incoming media, and British newspapers circulated to all intents and purposes unhindered. The BBC and other radio stations (including Vatican Radio and, of course, German broadcasts in English and Irish) were not jammed. The US embassy freely circulated its *Letter from America* to Irish opinion-formers, giving, after US entry into the war, a pro-Allied version of events, the details of which had in many cases been rigorously excluded from the Irish print and broadcast media. Reports sent abroad by Irish journalists were not subject to censorship until 1941; and the censor's job was in any case made considerably easier because no Dublin newspaper maintained war correspondents in the European theatre during the hostilities. As a result of this, Irish readers were left in a considerably more ignorant state than, for instance, their contemporaries in neutral Sweden and Switzerland (Ó Drisceoil, 1996b: 47).

The core rationale behind the censorship was expressed in a January 1940 memorandum by Frank Aiken, the minister for co-ordination of defensive measures. In this, he faced up squarely to:

> some self-styled democrats who would hold onto the peace-time liberalistic trimmings of democracy while the fundamental basis of democracy was being swept from under their feet by foreign or domestic enemies of their democratic state. Wise men, however, discard these trimmings when necessary in order successfully to maintain the fundamental rights of the citizens freely to choose by whom they shall be governed. (NAI, Taoiseach: 11586)

As censor, Joseph Connolly's somewhat choleric temperament, and his general impatience with what he saw as the pro-British proclivities of the press (especially the *Irish Times*), were evidenced in a November 1940 memorandum complaining that his censorship powers were inadequate:

> For example, there is at the present time no express power to compel a newspaper to handle news in a particular way, that is to say, to prescribe the part of, or the space in the paper, and the manner or form in which a particular item must appear ... Again, it is worth drawing attention to the fact that the Censorship is restricted to telling the papers what they must not publish – it has no power to insist on the publication of any particular matter. (NAI, Taoiseach: 10829)

Nonetheless his prohibitions were extraordinarily wide-ranging. One of Connolly's directives, for example, prevented the newspapers from publishing:

> matter which would or might be calculated to impair the financial stability of the State and in particular (a) matter reflecting adversely on

the solvency of the exchequer; (b) statements or suggestions tending or likely to cause uneasiness or panic among depositors in banks; (c) reports or rumours of the intended imposition of new taxation; (d) matter likely to provoke discontent among servants of the State or that would be liable to cause a withdrawal of labour by any branch of the civil service. (NAI, Taoiseach: S 11586)

The big stick with which the government had armed itself was reserved for persistent offenders, notably the *Irish Times*, but also papers like the *Kilkenny People*, a weekly provincial newspaper whose editor, E.T. Keane, was at least as truculent as Smyllie in his dealings with the censorship authorities (Hannigan, 1993: 15). The *Cork Examiner* got off relatively lightly; it was physically impossible, for all practical purposes, for that paper to submit proofs prior to publication as the other daily papers were required to do, and its conservative management operated a type of self-censorship under the censor's benign but distant supervision.

There was a debate on press censorship in the Seanad on 4 December 1940, but it fell rather flat, possibly because the worst excesses of the censorship had not yet manifested themselves. It was notable, however, for Aiken's public enunciation of the hard line earlier expressed in private:

> It is no change to the ordinary people of the country to have myself and my staff acting as Censor. It is only a change of function from the editor of the *Irish Times* to myself and my staff. It is we who have the final OK on what to cut out instead of the editors and in certain circumstances, when this country is in such grave danger as it is at the moment, I think the people are better pleased that the censorship should be in the hands of somebody who is responsible to the parliament elected by the people rather than of persons who are appointed by boards of directors and have no responsibility to the people, though normally they act in a decent way towards them. (SD, Vol. 24, 2610)

Connolly, and Michael Knightly, who succeeded Connolly in late 1941, played a cat-and-mouse game with Smyllie for the duration of the war. References to any of the numerous Irishmen serving in the Allied forces were invariably deleted by the censor, encouraging Smyllie to devise a range of circumlocutions. In December 1941, when the *Prince of Wales* was sunk off Singapore, Smyllie managed to convey to his readers critical information

about one of its crew – the *Irish Times* journalist Johnny Robinson, who had
left the paper to join the navy at the outbreak of the war:

> The many friends in Dublin of Mr John A. Robinson, who was involved
> in a recent boating accident, will be pleased to learn that he is alive and
> well. He is a particularly good swimmer, and it is possible that he owes
> his life to this accomplishment. (Gray, 1991: 153–4)

Although the *Irish Times*, and the Protestant-owned *Evening Mail*, suffered
substantially at the censors' hands (the *Irish Times* was forced to submit to
complete pre-publication censorship from early in 1942), it would be
incorrect to assume that other papers necessarily had an easier time of it. As
Donal Ó Drisceoil points out in his definitive study of this topic, the editor
of the *Irish Independent*, Frank Geary, was frequently as exasperated by the
censors' activities as his counterpart in the *Irish Times*. One censor wrote to
another in late 1942: 'It has always been our strength that Smyllie and Geary
cannot stand each other and I don't want to force them into an unholy
alliance if I can possible avoid it' (Ó Drisceoil, 1996a: 164). The *Sunday
Independent* also suffered, while the *Irish Press* concealed its grievances, even to
the extent of loyally supporting the government's censorship policy in an
editorial at the end of the war.

This sparring concealed a number of more serious issues for the press.
One was how to maintain freedom of speech for members of parliament and,
at election times, for parliamentary candidates. Chief among these – in both
categories – was James Dillon, son of an old Irish Parliamentary Party
politician, and a forthright advocate of the Allied cause, who resigned from
Fine Gael early in the war because he could not support the policy of
neutrality to which all Irish parties had committed. He became 'the most
censored of all politicians, with heavy treatment given to his letters to the
press and reports of his speeches' (Ó Drisceoil, 1996a: 264), regardless of
whether they were made inside or outside parliament.

The censorship of Dillon's speeches was at least overt. The censorship
authorities also resorted to discreet pressure, including, for example, visits by
members of the Gardaí to bookshops stocking offending publications.
Connolly outlined the success of this policy in a memorandum in 1941:

> There was a natural reluctance to alarm or antagonise either Dáil or
> public by taking what might be regarded as excessive powers and, accord-

ingly, an effort was made to secure the desired result as 'cheaply' as possible, and by indirect methods rather than full frontal attack. By this and similar devices we have up to the present achieved 90% of our objectives effectively, and often secretly. (NAI, Taoiseach: S 11586)

The occasional casualties of this effective and secret policy included publications from across the ideological spectrum, such as the *Leader*, an independent weekly that combined enthusiasm for Gaelic culture with scorn for many of the official policies designed to protect it, and which had speculated unguardedly about the possible advantages to Ireland of a German victory. Another was the *Irish Workers' Weekly*, an organ of the minuscule Irish Communist organization. The *Torch*, a Labour Party publication that was frequently critical (or would have been if passed by the censor) of the government's economic policy and its effect on working-class voters, frequently fell foul of the censor, as did the oddly named *Penapa*, a short-lived pro-Nazi paper that appeared in 1941.

One of the problems facing the censors – and which the censorship operation paradoxically exacerbated rather than moderated – was 'the continuing insistence in many circles on viewing all oppression through the lens of the British record in Ireland' (Ó Drisceoil, 1996b: 46). In this context, the ruthless suppression by the censorship of all atrocity stories meant that such rumours as filtered through to the public – such as those relating to the fate of Europe's Jews under the Nazis – were largely dismissed as British propaganda. The *Irish Press* expressed the view openly in an editorial on 1 April 1943: 'There is no kind of oppression visited on any minority in Europe which the Six County nationalists have not also endured' (Lee, 1989: 266). In the same year, all details of the massacre in the Katyn forest in Poland, in which 22,000 Poles were murdered by Soviet forces, were rigorously suppressed. Aiken, the minister responsible, explained this course of action many years later, telling an interviewer: 'What was going in the [concentration] camps was pretty well known to us early on, but the Russians were as bad – you have only to look at what happened in Katyn forest. There are photographs to prove that' (Ó Drisceoil, 1996b: 49).

The almost pathological even-handedness of the censorship left some Irish readers – at least those unable to access British or other radio broadcasts – quite unprepared for the horrors that were only fully revealed at the end of the war. Although it was theoretically possible for some Irish citizens to obtain less-censored war news from UK newspapers, a shortage of newsprint, deteriorating public transport and other factors had led to a decline by as much as 60 per cent in the circulations of Irish newspapers between the late 1930s and 1941, and even when the facts of the German genocide unfolded, many Irish readers continued to suspect that these stories were British propaganda (Ó Drisceoil, 1996b: 50). Discounting the inevitable benefit conferred by hindsight, it is inarguable that the confusion between military and moral neutrality embodied in the censorship practices of the day was, in its worst excesses, completely unjustifiable.

Journalism was never entirely cowed by wartime exigencies; indeed, a certain spirit of self-reliance engendered by the war encouraged two initiatives that illuminated the media landscape. One was the creation of *The Bell* in October 1940 by the writer Seán Ó Faoláin, who edited it until 1946. *The Bell* acted as a powerful and effective focus for the scattered forces of literary and social liberalism in the country at the time: in its first editorial, it proclaimed itself a journal of the present and the future, not of the past:

> *The Bell* is quite clear about certain practical things and will, from time to time, deal with them – the Language, Partition, Education, and so forth. In general *The Bell* stands, in all such questions, for Life before any abstraction, in whatever magnificent words it may clothe itself. For we eschew abstractions, and will have nothing to do with generalisations that are not capable of proof by concrete experience. Generalisation (to make one) is like prophecy, the most egregious form of error, and abstractions are the luxury of people who enjoy befuddling themselves methodically. We prefer, likewise, the positive to the negative, the creative to the destructive. We ban only lunatics and sour-bellies. (*The Bell*, 1 (1): 8–9)

The Bell's virility even prompted the wartime Fianna Fáil minister for finance, Sean MacEntee (who was given to writing somewhat mannered lyric poetry in his spare time), to respond favourably in 1942 when Ó Faoláin approached the government for help. MacEntee was only persuaded against this when colleagues pointed out that if a subsidized publication attacked the government, the government could not withdraw the subsidy without being accused of censorship (UCD, MacEntee: P67/246). In its willingness to challenge established hegemonies, especially that of the Catholic church, and its embrace of cultural and political variety, it played a major role in the emergence not only of a newly vigorous Irish literary tradition, but of a willingness to imagine alternatives that contributed to the defeat of the Fianna Fáil government, after sixteen years of continuous rule, in 1948.

Another exemplar of this new tendency was the emergence of *Comhar* ('*Co-operation*') in May 1942. This monthly publication in Irish was the offspring of the Comhchaidrimh, an association of university societies dedicated to the promotion of Irish language and culture, which was founded in 1935. It maintained a consistently high standard of literary and political journalism, acting as a nursery for a number of activists who later became better known in national journalism and politics (Nic Pháidín, 1987: 71). Its first issue carried an article by the youthful Conor Cruise O'Brien, defending his alma mater, Trinity College Dublin, against charges of anti-Irishness. Contributors were to include the writers Frank O'Connor, Seán Ó Faoláin, Brendan Behan, and Myles na gCopaleen (the civil servant Brian O'Nolan) (Nic Pháidín, 1982). O'Nolan had, since 1941, been afforded space in the *Irish Times* by Smyllie for an irregular, multilingual column (Irish, English, Latin and Greek were among the languages employed) that poked devastating, literate fun at Irish pretensions of every kind. An equally independent spirit led to the foundation in 1943 of *Inniú* ('*Today*'), an idiosyncratic weekly newspaper founded by O'Nolan's brother Ciarán O Nualláin. It was written in Dublin but – in order to help the economy of the underdeveloped western areas – printed in Co. Mayo. During the war it was plagued by transport difficulties, and also by a well-intentioned but disastrous scheme aimed at boosting circulation by offering to correct essays in Irish by schoolchildren. This produced a huge workload, but very little finance, as the schools were slow to settle their bills.

The position of broadcasting in the media system was somewhat anomalous. Although it was carried out by a government department, rather than by an independent commercial organization, Radio Éireann's relationship with the censorship operation was at first poorly defined. Its news operation incurred official displeasure after it occasionally broadcast items already censored from the print media. Eventually, given the extreme sensitivity of the situation (its broadcasts could be heard outside Ireland, where the Irish print media were invisible), a regime was established whereby news broadcasts had to be read over in advance to the head of the government information bureau, Frank Gallagher, and on occasion to de Valera himself.

For all its deficiencies as a conduit for news, however, Radio Éireann's reputation was enhanced thanks to an unusually sophisticated minister, P.J.

Little, a devotee of classical music. Under his stewardship, the station's musical output increased, both in quantity and in quality. The rest of its output consisted largely of well-established favourites, some borrowed from BBC formats such as *Question Time,* and drama. A number of initiatives were also taken to encourage the speaking of the Irish language, but any outreach in the most obvious direction – the schools – was stymied by the wartime difficulties in obtaining and repairing radio sets and batteries, as well as by the conservatism of the religious managers of the country's church-owned primary schools. Little was also responsible for the re-establishment in 1944 of the Broadcasting Advisory Committee, which had had a fitful existence up to 1933, and which had generally been regarded as a nuisance by the civil servants in control of broadcasting. Its second incarnation lasted for a full decade (Gorham, 1967: 117–35).

The communications difficulties engendered by the war had a major impact on broadcasting in one sense: the country's isolation, and the perceived need to defend its neutral stance in every possible forum, led to the re-examination of an earlier project to establish a short-wave radio service. This had initially been proposed in the 1920s by Marconi and other operators anxious to break the British post office and cable operators' monopoly on transatlantic communications. Domestic financial stringencies, and political caution, meant that these proposals did not receive serious consideration, but the change of government in 1932, and the simultaneous launch of the BBC's Empire Service (later World Service), prompted a reconsideration, not least by the Broadcasting Advisory Committee, which proposed in July 1932 that a short-wave station be established for the benefit of 'the 4.54 million Irish people living abroad' (NAI, Communications: TW 8828). It was 1937 before pressure from emigrants (and Irish newspapers) saw an interdepartmental committee set up to study the possibilities (Cullen, 1991: 13). No doubt hastened by the impending hostilities, installation of equipment for the new station near the existing medium-wave national station at Athlone was completed in February 1939, and the first short-wave broadcast was made on St Patrick's Day 1940.

Initial results were poor. The transmitter was underpowered, and reception reports from North America were sporadic. Increasingly desperate attempts were made by a committee that included Professor E.T.S. Walton of Trinity College Dublin (who would in 1951 win the Nobel Prize in Physics) to improvise equipment that would boost the signal strength, but all ended in failure, and the attempts were effectively abandoned in September 1944. The equipment was recommissioned in May 1945 for a rebroadcast, in a French translation, of de Valera's reply to Churchill's VE day attack on Irish neutrality, but this was largely symbolic.

The renewed availability of broadcast equipment after the war rekindled the government's enthusiasm, and under P.J. Little, planning proceeded apace. A new high-power transmitter was acquired, a forest of antennae covering 40 acres was erected, and – most importantly of all – government approval was secured for a huge level of expenditure, not only on equipment but on actually making programmes.

In April 1946, Little announced to parliament a programme of enormous expansion for Radio Éireann. An increase in expenditure of almost 60 per cent over the previous year was to be devoted to enlargement of the existing orchestra, and the creation of a second orchestra, new positions for outside broadcast officers and a repertory drama company, whose players formed the backbone of radio drama thereafter. News coverage was strengthened, and staff scriptwriters and features producers were hired. Although the talents thus employed would be utilized across both the medium-wave domestic service and the proposed revived short-wave service, without the impetus of the short-wave plan few, if any of the improvements would have been made. Their primary purpose, as Little announced on radio, was to present the 'everyday story of the new Ireland, spoken with its own voice' (*IP*, 20 Feb. 1947).

It was a false dawn, but for reasons none could have anticipated. The February 1948 general election saw Fianna Fáil replaced by a new government convinced of the financial profligacy of its predecessor. It could not rescind the staffing improvements but it could, and did, put the short wave service into cold storage.

The political division between Northern Ireland and the Irish Free State (which did not describe itself formally as a republic until 1949) was echoed in the media on either side of the border. Few publications crossed from North to South, although the main Dublin newspapers had modest circulations in the North, particularly in Belfast. This did not mean that Northern papers were insensitive to political developments in the South; de Valera's election victory in 1932 had been greeted by the *Northern Whig* with an editorial (on 13 July) declaring that:

> the destinies of the South today are in the hands of a coterie of Anglophobe agitators with no understanding or experience of government who are driven forward in their extremist policy by a secret

organization who seem quite prepared to go to any length to bring
Ulster under the control of the Free State. (Kennedy, 1988: 145–6)

The *Irish News*, for its part, greeted de Valera's 1932 election with
unrestrained enthusiasm. Later the same year, in a mirror image of what
happened to the *Irish Press* in the dying days of the Cosgrave government in
Dublin, it found itself in the Belfast courts, prosecuted under the Special
Powers Act for publishing a confidential circular issued to Orangemen, which
the prosecution alleged was prejudicial to peace and order. Even in these
unpromising circumstances, however, unionism occasionally appeared to be
less of a monolith than was generally supposed. A small group of liberal
unionists, with links to Queen's University, coalesced in the late 1930s around
a journal modestly entitled the *New Northman*, which criticized the Special
Powers Act and 'the junta of self-seeking politicians who call themselves
Ulster' (Kennedy, 1988: 232). They were later to form the nucleus of the Irish
Association for Cultural, Economic and Social Relations.

Unlike Dublin newspapers, the *Telegraph* had its own war correspondents –
one assigned to each branch of the services. Like all papers, North and South,
it was subject to censorship, but the activities of the censorship office in
Belfast, operated by the minister of information – by some reports draconian
(Brodie, 1995: 69) – have not been as well documented. Censorship was not
an issue at BBC NI, in large part because regional broadcasting from Belfast
was suspended for the duration of the war.

The dearth of censorable material did not mean that the Belfast censorship
office ceased to exist. The problem, however, was that unionist sentiment and
British government media policies were pulling in opposite directions
(Cathcart, 1984: 118–27). London felt that promoting Belfast-Dublin co-
operation might pressure Ireland to abandon neutrality, whereas Belfast felt
that any cross border co-operation was little short of treason.

The restoration of regional broadcasting from Belfast in July 1945 did not
lessen the political pressure from unionism. Basil Brooke, the Northern
Ireland prime minister, proved as eager as his predecessor, J.M. Andrews, to
interfere with broadcasting, either by making representations to London, or
by making clear, as he did in April 1946, his belief that BBC NI should be
controlled by Stormont (Cathcart, 1984: 131). The BBC managed to fend off
these attempts without much difficulty, but the price was paid internally in
Northern Ireland, where open discussion of political issues was still
conspicuous by its absence.

For all the differences between the two parts of the island, some organizations concerned with journalism as a profession had for many years operated on both sides of the border. These were the Institute of Journalists and the National Union of Journalists, both of which were effectively British organizations with pre-1922 footholds in Ireland. The IOJ, founded in 1884, was first in the field, and included in its membership proprietors as well as employees of newspapers. The NUJ was founded in 1907 for employees only, and was much more of a trade union; it established its first presence in Ireland in 1909.

The same year saw the first of a number of attempts to organize a separate journalists' organization in Ireland. This was the Irish Journalists Association, whose weakness was evidenced by the fact that it did not publish its first newsletter until October 1914 (NLI, IR 07 g I). In October 1915, representatives of the NUJ had discussions in Dublin with IJA members about a possible merger, but IJA members resisted the adoption of trade-union status. By the time the NUJ formed its first formal branch in Dublin in 1926, the IJA had to all intents and purposes ceased to exist. The new NUJ branch quickly achieved a membership of about eighty, mainly from the national papers. A branch was formed in Belfast with seventeen members. The foundation of the *Irish Press* in 1931 provided a notable boost, as employees of that newspaper were strongly trade-unionist in outlook, and rapidly achieved a virtual closed shop there (Bundock, 1957: 110).

Growing union militancy during the war years provoked a number of developments on either side of the border. In the South, the union faced a crisis decision following the passage of the 1941 Trade Union Act, requiring all unions to register with the government and pay a large fee before being entitled to negotiate on behalf of their members. In the UK, the NUJ central organization was hostile to such a development, objecting to any government interference with trade-union rights. Accordingly, it urged its Irish members to set up a separate organization, and offered to pay the £1,000 required for a negotiating licence, as well as providing the new organization with grants in each of the following two years. This proposal, however, was defeated at the union's annual delegate conference in Leeds in 1942 after the Dublin-based author and journalist R.M. Fox argued that it amounted to abandonment of the Dublin branch. Later that year the union reluctantly complied with the registration procedures, while putting on record its determination 'to resist by every means in its power the application of these dangerous principles to the trade union movement' (Bundock, 1957: 162).

In Belfast, there were problems with the employers, who revived the Belfast Newspaper Society, a formerly dormant proprietorial association, and rapidly came to an agreement on pay with the IOJ. By 1947, the NUJ's Belfast and district branch – an enlargement of the original Belfast branch – acquired members from the editorial staffs of all the newspapers, and approached managements (initially the *Belfast Telegraph*) for recognition. The proprietors, adopting a joint position, refused. The *Telegraph* decreed that 'no organisation, social or otherwise, can operate or function in any part of the company's premises unless with the direct and specific permission of the directors'. Any requests to meet management, they insisted, could come only through department heads (Brodie, 1995: 93–4). In Dublin, the union was stronger and, in 1947, negotiated the first of a series of agreements that resulted in substantial improvements in pay and conditions (Bundock, 1957: 223). Although this contributed in turn to the growth of the union, it was to be almost two decades before the Irish organization had its own administrative structure, and indeed professional unity was repeatedly threatened by controversies in the 1950s and 1980s.

Apprehension was great in Dublin, driven partly by a communist 'scare' launched by Fianna Fáil in the course of the general election in February 1948. The American legation in Dublin was particularly concerned about this, forwarding reports to Washington about communist activity in Ireland and, in particular, rumoured communist infiltration of the *Irish Press* and the National Union of Journalists. Some of these rumours reached de Valera himself, who was concerned enough to seek reassurance from his editor, William Sweetman, about them, and in particular about the political loyalties of the *Irish Press*' foreign editor, Sean O'Neill, whom the Americans thought 'systematically takes advantage of his position to slant the news ... minimises anti-Soviet statements and uses a series of journalistic devices to misrepresent the news as originally received' (USNA, 841D.00B/9–2349, 3 Oct. 1949).

It was hardly coincidental that a serious attempt was made at about this time to set up a wholly Ireland-based union for journalists, but the fear of communist infiltration was a minor motivation. More significant in the long term was the desire of the post-war government to promote industrial organization, and its hostility towards UK-based unions generally. This had already been made manifest in a major split in the trade-union movement, which now had one organization for Britain-based trade unions, and another for purely Irish organizations. This general split now spread to journalism: in 1949 a number of journalists, with the active support of the Irish Transport and

General Workers' Union, set up a new organization known as the Guild of Irish Journalists, and issued its first press statement on 22 March, asking bluntly, 'Why should Irish journalists be controlled from London?' (Horgan, 1997: 144). Like its predecessor, the Irish Journalists Association, it eventually became comatose, although it experienced a brief revival in markedly different circumstances more than thirty years later.

Events in Dublin continued to cast a long shadow northwards. Sometimes it was benign: the post-war developments at Radio Éireann may well have played a role in encouraging the creation of the BBC NI orchestra in 1949. But more usually it was a cause for concern, and apprehension. This was particularly so when the constitutional issue raised its head, as it did in a marked manner after the election of a new, non-Fianna Fáil government in Dublin in February 1948. When the BBC World Service broadcast extracts from an irredentist speech by the new Dublin taoiseach, John A. Costello, a Westminster Unionist MP immediately tabled a motion deploring the rebroadcasting of 'passages offensive to Northern Ireland' in Costello's speech. Ironically, Costello was almost equally offended, as the BBC had judiciously excised some of the more inflammatory of his statements about Irish unity (Cathcart, 1984: 145). This perceived slight was to have an unforeseen consequence: the revival of the plan for a high-power short-wave service in Athlone, which Costello's government had killed off. In the South as well as in the North – and with notably more success, given the organic nature of the relationship between government and broadcasting – interventionist media strategies were still a popular option.

4 Coming of age, 1947–57

The end of World War II was not accompanied by the rapid upsurge in living standards many had confidently expected. One of the major unanswered questions in Irish economic history is why Ireland did not benefit, to the extent apparent elsewhere, from the economic growth that accompanied reconstruction in Europe. De Valera's grip on power was, partly as a result, beginning to weaken, and by the end of 1947 he was contemplating a snap election to catch his opponents unawares.

As part of his preparation for this contest, he partially addressed the claims that his government was not doing enough for the Irish language by subsidizing Irish-language periodicals for the first time. A grants system devised by his minister for finance, Frank Aiken, put a number of key publications on a much sounder financial basis: *Inniú* received £4,900; *Comhar* got £1,150, with a further £750 if it was printed in Dublin; and *Feasta*, which was launched in 1948 by Conradh na Gaeilge (the Gaelic League) to replace its more modest journal *Glór* (*'Voice'*), received £1,800.

These measures were insufficient to stave off defeat. Fianna Fáil's ejection from government for the first time in sixteen years had rapid consequences for Irish journalism. The party moved almost immediately to establish the *Sunday Press.* The Irish appetite for Sunday newspapers has always been substantial, and for years the proportion of British newspapers sold in the country on Sundays has been substantially higher than for the daily press on weekdays (Horgan, 1993b). Planning for the paper began in 1948, and the first edition was published on 4 September 1949. Initially the project encountered problems similar to those its morning stablemate had overcome: the Independent group – whose *Sunday Independent* was of course threatened by the newcomer – let newsagents know that if they stocked the new Sunday paper, supplies of the *Sunday Independent* would be withheld.

The *Irish Press'* managing director, Sean Lemass (translated to this position from his former role as cabinet minister, and still a member of parliament), had to devise alternative strategies. These included approaching the proprietors of small shops in many rural towns and villages, who had never sold newspapers before, but who were known to be loyal to the party and who would be prepared to open on Sunday to stock the new paper. In this way, a parallel distribution network was successfully organized. Rather more unusually, Catholic priests favourable to the cause were also invited to volunteer, and many did so, some agreeing to transport bundles of the newspaper in their cars to the churches, so that the crowds emerging from Mass could buy them immediately afterwards (Oram, 1983: 239). The enterprise was immediately successful – peaking at over 400,000 weekly sales in the 1960s, but not at the expense of the *Sunday Independent*, suggesting that many of its readers had either switched from a UK Sunday paper or might not have bought a Sunday paper of any description before.

As Irish journalists revelled in the opportunities offered by the emergence of a new national paper, and the major political changes that were taking place, they were also faced by what many considered unwarranted government interference in the sphere proper to journalism. This was the government's plan to establish a new body, the Irish News Agency, staffed by journalists but controlled by the government, which would have an important role vis-à-vis foreign (principally British and American) publications.

This development had its origins in three major factors. One was political – the desire of the political party Clann na Poblachta and its leader, the former revolutionary Sean MacBride, to bring international pressure to bear on Britain to end partition. The second was the sense that the Irish case on partition had never been properly explained to the outside world, as virtually all news from Ireland was filtered through British news agencies before reaching the international media. The third was a growing dissatisfaction with the work of the government information bureau. Set up in 1934, the GIB's role had always been ill-defined. Originally established to ensure that all ministerial press statements would be issued via a channel controlled by the head of government – thus avoiding embarrassing departures from government policy – a succession of strong cabinet ministers had ensured that it never really fulfilled this role.

MacBride's belief that Ireland was surrounded by a paper wall that prevented Irish grievances against Britain from obtaining an international airing had some basis in fact. All the major international news agencies had correspondents in Dublin (MacBride had once worked for the French agency Havas), but their reports were generally routed through London desk editors unsympathetic to Irish irredentism, who assumed international audiences would have little interest in it (Horgan, 1993a: 33). The possibility that the world at large – outside the large Irish communities in Britain and the US – might be relatively indifferent to Ireland's historic grievances did not prevent MacBride from persuading the government to set up the Irish News Agency.

Early newspaper reports suggested that the agency's work would be confined to sending out articles by airmail, and to the use of short-wave radio, but as more details emerged, journalists became openly hostile. The *Irish Independent* suggested that the plan 'smack[ed] too much of a plan to set up a propaganda department' (*II*, 17 June 1949), and the *Irish Times* expressed the view that 'Mr MacBride's diminutive agency certainly will not galvanise the apathetic audiences of the great world into a state of appreciation' (*IT*, 15 July 1949).

A lot of this opposition was substantially motivated by self-interest. This was openly admitted: an opposition member of parliament, himself a journalist, told the Dáil that he was speaking on behalf of working journalists, 'who feel that their livelihood is being attacked in this bill', and that the agency 'will be in competition with other journalists who send "copy" out of this country' (DD, 13 July 1949). MacBride responded with an undertaking that the INA would not deal in 'hot' news. This created a new problem: how could an organization that did not distribute news locally be legitimately described as a news agency?

Ill-starred though its birth may have been, the new agency was given considerable resources and housed in MacBride's department of external affairs, where most of its day-to-day business was overseen by its managing director, Conor Cruise O'Brien, then working in that department as a civil servant. One of the first decisions of its board was to jettison MacBride's pledge that it would not deal in 'hot' news. This followed a proposal from one of MacBride's close associates, Noel Hartnett, who was a member of the board. Other board members assumed that this was being done with MacBride's covert approval, but in fact the relationship between MacBride and Hartnett had soured. Thus Hartnett's proposal was made in the full knowledge that it would embarrass the minister, as indeed it did.

The agency nonetheless got into its stride quickly, and within a year was generating about 9,000 words a day (Horgan, 1993a: 35). Its staff, which was to rise to almost 50 by 1952, included many of the country's brightest young journalists, who were paid at rather better rates than those offered by the national newspapers. It secured an agreement with the Hearst-controlled INS news service in the United States, whereby INS carried INA-generated stories for its American clients, while the INA had a monopoly on selling INS material (principally photographs) to the Irish media.

The more it succeeded in selling Irish news abroad, however, the more it aroused opposition among Irish journalists. It also fell foul of public sentiment when, as occasionally happened, it exported news appearing to show Ireland in a bad light. It combined attempts to become financially self-sufficient by selling news, with a less-publicized attention to the propaganda role MacBride had originally envisaged for it. This uneasy dual role was defined by O'Brien in November 1950 in a note to the Irish government's public relations officer in New York, when he said that the agency:

> will function as a genuine news agency and not as a propaganda body. I must add for your own information that this will not prevent the agency from giving news and stories about Partition. This, however, will be thoroughly reliable and discreetly administered. (NAI 5/340/12/88)

The agency's (unpublished) report for 1951–2 made much of its success in distributing to a claimed audience of 8.5 million people a photograph of policemen in Northern Ireland batoning unarmed nationalist demonstrators in Derry. This, the agency proudly informed its political masters, was:

> probably the first time that a highly important political fact – viz that partition is maintained by force – has been so effectively brought across to an audience of large dimensions. The quasi-monopoly of the handling of Irish news for the world's press, once possessed by two or three of the great international news agencies, is now broken. Secondly ... the state of affairs whereby all Irish news was edited in London before issue ... is now a thing of the past. (NAI, Taoiseach: 14544 B)

This was one of a number of INA attempts to avert attempts by the department of finance to close it down because it was consuming ever-increasing quantities of public funds. Its revenue-generating capacities had been seriously impaired when, at the end of 1951, following NUJ pressure, de Valera, taoiseach again after three years in opposition, agreed that its distribution of 'hot' news would cease.

The journalists who clipped the fledgling agency's wings were, the department of external affairs observed tartly, in 1953:

> in large measure not engaged in the active collection of news. At least 90 per cent hold editorial positions on the Dublin daily newspapers and recast the news submitted to those newspapers by working journalists who receive no payment for the exported material. It is easy to see why such vested interests should be continuously and actively vocal in their

opposition to a national news agency which pays cash for all news written. (NAI, Taoiseach: S 14544 B)

At the time this was written, the *Irish Independent* alone had among its staff the Dublin correspondents of the Press Association, Reuters, the Associated Press, the Exchange Telegraph financial newswire and a Canadian news agency.

The INA limped along for a further three years before being finally closed down in May 1957. For all its faults, the agency was probably one of the most exciting developments in Irish media in a period otherwise marked by the regular expression of a crude political and cultural isolationism. As one historian commented (although his remarks are somewhat unfair, particularly to publications like *The Bell*):

> Regrettably most Irish journalism in this period [1932–58] had contented itself with the reportage of events and the propagandist reiteration of the familiar terms of Irish political and cultural debate until these categories became mere counters and slogans often remote from many actualities. Irish journalism therefore comfortably reinforced the prevailing sense that Ireland, marked as the nationalists constantly stressed by distinctive social, religious and linguistic forms, was somehow different from the rest of the world. It did not challenge Irishmen and women to reflect seriously on their own reality. (Brown, 1981: 204–5)

From early in 1950, the government was deluged with a series of resolutions passed by local authorities in different parts of the country calling for action against imported publications (NAI, Taoiseach: S 2919B). Leitrim county council, for example, demanded a prohibition on the importation of Sunday newspapers 'in view of the offensive and obscene nature of articles continually appearing therein'; in April, Galway corporation followed suit, charging that these publications were 'detrimental to the moral life, national aspirations and economy of the country'; in the same month, the Portlaoise town commissioners alleged that these publications were 'killing the soul of the nation'. The taoiseach, John A. Costello, demurred. Proposals for banning newspapers were, he said, a purely negative approach: 'the only way in which you can deal with matters of that kind is by raising the standard of taste of the people' (DD, 24 Apr. 1951).

In spite of conservative attitudes like those that prompted local authorities to propose bans, the success of *The Bell* had shown that there was a constituency for the venturesome, even if the implied invitation was effectively taken up only by the *Irish Times*, where Smyllie continued as editor until his death in 1954. In July 1949 the paper published a series of articles on venereal disease, which were 'greeted with dismay in higher Catholic circles, and the subject was almost completely boycotted by the Catholic press' (Blanshard, 1953: 152). The following year, it became the venue for an extraordinary correspondence under the generic heading 'The Liberal Ethic', later published by the newspaper itself as a pamphlet, in which a wide range of literary and political figures elaborated, defended or attacked the ideology to which the paper was increasingly explicitly committed.

This found its most dramatic expression in April 1951, amid the 'mother and child' controversy, a government crisis about a proposed health scheme to which the Catholic hierarchy and the medical profession were resolutely opposed. After intense and fruitless negotiations with these interests behind closed doors the minister for health, Dr Noel Browne, was forced to resign. His last act before leaving office was to release to the press a full dossier of the correspondence between himself and various bishops and politicians. The newspapers were initially hesitant, as publication was technically a breach of the Official Secrets Act, and therefore a criminal offence, but Smyllie decided to publish regardless: the other papers, thus encouraged, did the same (*IT*, 12 Apr. 1951). It was a remarkable breach, not only of the criminal law, but of the polite conventions of political journalism. Nor were these events ignored in Northern Ireland: the *Belfast News Letter*, long the unofficial organ of the Unionist Party, gleefully reprinted the documents as a booklet – 'Southern Ireland – church or state?' – which was available thereafter from the Unionist Party headquarters.

This was a brief blip. For the most part, the print media operated within boundaries that were probably permeable to an extent that is difficult to quantify. In March 1953, the circulations of the major papers were as follows: *Irish Independent*, 203,206; *Irish Press*, 198,784; *Irish Times*, 35,421; *Cork Examiner*, 45,917. The largest-selling Irish Sunday newspapers were the *Sunday Independent* (395,507), followed by the *Sunday Press* (378,454) (DD, 19 Mar. 1953).

The major problem faced by the *Irish Times* was the narrowness, and vulnerability, of its operations. Its presses were idle for most of every day, and completely on Saturdays. It did not publish circulation information, unlike the *Irish Independent*, which had been the first Irish national paper to submit returns to the Audit Bureau of Circulation. Indeed, its circulation figures were not even made available to the board until the 1950s, on the grounds that it was commercially sensitive information that might cause the paper damage if it leaked into the public domain. The board was a group of five Protestant businessmen, who controlled all the voting shares. When one of their number died or wished to resign – which happened infrequently – the remaining directors habitually purchased his shares, and redistributed them to the next board appointee. The fact that many businesses in Dublin were Protestant-owned ensured a steady supply of commercial advertising, to an extent probably not warranted by the paper's circulation; but these factors could not be expected to continue into the indefinite future.

It had only two other small publications. One was the weekly *Irish Field*, dedicated to the horse-racing and breeding industries, which was a profitable niche publication but which did not really demand much of the company's printing resources. The other was the *Times Pictorial*, which had been started in 1941. All of its rivals had, since 1954, three substantial papers to keep their presses turning and their income steady. To remedy its dangerous reliance on a single product, the *Irish Times* in 1957 started a new Sunday paper, the *Sunday Review*, and in 1960 bought an evening newspaper – the *Evening Mail*. The *Mail*, like the *Irish Times* Protestant-owned, was an old-fashioned broadsheet newspaper with a declining circulation. Under its new owners, it became a bright and aggressive tabloid, but the experiment was a costly failure. The old readers did not like their paper in its new guise and, before it could attract a viable circulation among the younger urban readership at which it was aimed, mounting losses forced its closure in 1962. Until then, Dublin, with three evening newspapers, had if anything been oversupplied.

Even before the closure of the *Mail*, the *Times Pictorial* had gone to the wall in 1958. The closure of its Sunday tabloid title, the *Sunday Review* in 1963, was an added blow. It suffered, like the *Evening Mail*, from the *Irish Times'* small capital base, and might well have survived and prospered if its parent organization had had deeper pockets. It was as different from its parent in style and content as it is possible to imagine. Its chief innovation was in the field of political journalism: its weekly feature 'Backbencher' evolved from a loose collection of anecdotes about politicians into an irreverent, hard-edged and demotic treatment of the political process, characterized by inside information of a type and quality that journalism in Ireland had not known until then. This was paralleled by, and dependent on, a loosening of the traditional disciplines that had kept cabinet leaks to a minimum: a new generation of younger politicians, particularly in Fianna Fáil, impatient at the slowness with which their elders were prepared to move aside, began to use the media, and in particular the 'Backbencher' column, to push their claims for advancement. On the closure of the *Review*, the column was translated successfully to the *Irish Times*. Since the beginning of the 1950s, therefore, the *Irish Times* portfolio of publications had expanded to no fewer than five publications: the successive closure, from 1958 onwards, of the *Times Pictorial*, the *Evening Mail* and the *Sunday Review* left it economically shaken and apprehensive about the future.

No such doubts afflicted the *Irish Independent* and its associated evening and Sunday newspapers, which went from strength to strength. Its ownership had been to some extent diluted by marriage, so that the Murphy family were now joined at board level by members of a Dublin medical family, the Chances. Still, it remained effectively a family business, whose smaller shareholders paid little attention to the politics of the company's board, content to receive their generally substantial dividends. In some areas, the *Irish Independent* had a virtual monopoly of advertising. Teaching vacancies in Catholic primary schools, for example, were invariably advertised in the daily paper by the parish priests who managed the schools.

To suggest, however, that the Independent group papers were closely linked to one political party would be an overstatement. They had been intermittently critical of the Free State government in the 1920s, and their initial support of Fine Gael owed more to the threat to the civic and political hegemony of the middle classes that it discerned in Fianna Fáil than to any inherent conviction about the correctness of Fine Gael policies. The Independent group's ongoing support of Fine Gael was primarily because the *Irish Independent* supported the Catholic middle class and commercial groupings, of which Fine Gael happened to be the most coherent and conservative political expression. Even this is to some extent a generalization: some of the group's editors had a fairly robust attitude towards, and maintained strong journalistic links with, the main figures in all the political parties.

There was one interesting anomaly in all of this. In the 1951 'mother and child' controversy, the young and impulsive minister for health, Dr Noel Browne, launched a scheme that would have provided free medical care for all mothers and for all children up to the age of 16. This scheme was bitterly opposed by the Irish medical profession, who enlisted the Catholic bishops ain their support. The *Independent*, which might have been expected to weigh in heavily behind the bishops and the medical profession, maintained an

unexpected editorial silence. Although it would not have been widely known
or remarked on at the time, the reason for this was in all probability the fact
that the Chance family, which was represented on the board, had some fifteen
years earlier informally adopted the young Noel Browne, then an orphan, and
paid for his university education. Ideology and practical politics dictated that
they could not support the stand he had taken; but neither would they allow
their papers to attack him.

As the controversy reached its climax, the *Irish Independent* sailed serenely on:
in the issue that published the dramatic correspondence, it also regaled its
loyal readers with the thirty-fourth in a fifty-part series of articles on 'The
lives of the popes'. This was very much in character. Its Sunday stablemate
provided a platform for the last article by the redoubtable Fr Richard Devane,
on the evil influence of crime comics (*SI*, 22 Mar. 1951): he was to die two
months later (*II*, 24 May 1951). Part of his legacy was the organization known
as Cosg ar Fhoilseachain Gallda ('Ban on Foreign Publications'), which, at a
mass meeting in Dublin in 1952, announced plans to establish committees in
every parish to boycott foreign (generally British) magazines and newspapers.
These committees were presumably responsible for the appearance in many
parts of Dublin, early in 1953, of posters declaring in heavy black type: 'THE
FOREIGN PRESS IS A NATIONAL MENACE' (Blanshard, 1953: 102).
The organization even secured a meeting with the taoiseach, Eamon de
Valera, to press its point of view on him, but without success (Adams, 1968:
164). De Valera maintained that all but the most objectionable foreign publications were best dealt with by maintaining the import tax introduced in 1932
(NAI, Finance: FO22/0021/33). The combined circulation of the three
national morning papers in 1953 was almost 450,000 (*IT*, 19 Mar. 1953),
although per capita distribution of these was still low – about a third that of
the UK, and less than half that of Australia, Norway, Denmark or Sweden
(*IT*, 19 Aug. 1953).

In 1954, as the *Irish Independent* prepared to celebrate its fiftieth year in
existence, it wrote to a number of senior Catholic churchmen, inviting them to
contribute special messages for an anniversary supplement. Churchmen of
other denominations do not appear to have been similarly invited, such was the
narrowness of the paper's cultural and social focus. The cardinal archbishop of
Armagh and primate of all Ireland, Dr d'Alton, told the paper that it could:

> justly claim that during the 50 years of its existence it has maintained a
> high standard of journalism and respected the decencies of life. It

appeals to a wide circle of readers because it is usually varied in content, interesting, informative, without exploiting the cheap or sensational. It has always endeavoured to promote the best interests of the nation. (*II*, 22 Dec. 1954)

D'Alton's counterpart in Dublin, the powerful Archbishop John Charles McQuaid, was equally to the point. The paper had been marked, he told the editor, Frank Geary, by 'your policy of distinctive loyalty towards the Church' (*II*, 31 Dec. 1954). Geary responded in an editorial on 3 January 1955, in terms that could well have been published fifty years earlier:

> In the future, as in the past, we shall endeavour to live up to our title, to be both Irish and independent, allied to no party, free to criticise or to help any or all as the interests of the nation may demand, using the mechanism of progressive newspaper production and the power of honest, sober journalists to enlighten, educate and serve the Irish people.

This period marked the high point of conservatism insofar as the *Irish Independent's* journalism was concerned; and it was echoed in other commercially owned publications such as the weekly *Standard* and *Irish Catholic*. Neither of these publications had a circulation even remotely resembling that of the secular papers. They were 'strong points on an ideological frontier, and the average Irishman, not feeling the frontier to be menaced, nor much desiring to put it forward, is not particularly interested in them' (O'Donnell, 1945b: 38–9). But even these publications had their differences: on the occasion of the abdication of Edward VIII in 1936, the *Standard* confined itself to a few well-chosen phrases about English immorality. The *Irish Catholic*, however, printed on its front page an article that made this the occasion for advancing the claim to the English crown of no less a personage than the (Catholic) Rupert I of Bavaria. Practically unnoticed by such trenchant defenders of the faith, however, there were stirrings in the clerical undergrowth – more particularly in the field of religious magazines, hitherto a repository of ultra-orthodox thinking and writing (Adams, 1968: 65–8). There was no official church-owned Catholic newspaper, although the Catholic hierarchy had entertained the idea of starting such a paper in 1927, eventually deciding to support the foundation of the lay-owned *Standard* rather than initiating a publication under its direct control. The Church of Ireland had the *Church of*

Ireland Gazette (founded in 1856), the Presbyterians had the *Presbyterian Herald* and the Methodists had the *Methodist Newsletter*.

The generally obsequious approach of the mainstream media, notably the *Irish Independent*, towards the Catholic church, had made the creation of a church-owned newspaper unnecessary. However, change was on the way. In 1953, this had been noticed even by as fierce a critic as the American Paul Blanshard, who wrote that:

> Occasionally in clerical journals one can find among the writings of the younger priests a few faint intimations of the love of freedom, a few oblique suggestions that they are suffering pain and humiliation because they have been asked in an age of democracy to dedicate their lives to a system of power in which they have no voice or vote. (1953: 69)

Although he did not name it, Blanshard was probably referring to *The Furrow*, a monthly pastoral journal launched in Maynooth, the country's largest Catholic seminary, in February 1950. The association with Maynooth, long regarded as the seat of episcopal power in Ireland, could be misleading: the journal's progenitor, the Revd J.G. McGarry, was a pastoral theologian who realised that the Catholic church in Ireland needed a fresh new voice on religious and theological affairs.

On the other side of the river Liffey from the *Irish Independent*, the *Irish Press* was also consolidating and expanding. Having lost power in the May 1954 general election, Fianna Fáil launched yet another newspaper later the same year, the *Evening Press*.

The *Evening Press* shook up the newspaper business in Dublin considerably. Its news-hungry editorial style was paralleled in the commercial department, which canvassed advertisements – particularly small ads, the staple of the evening newspapers for many years – from advertisers in other papers, notably the *Evening Herald*. It bought a fleet of motor-scooters, which delivered papers to the outlying Dublin suburbs much faster than the other papers' vans, and it pioneered the 'bush' system – a method of adding late news to previously blank spaces on the front and back pages of the paper at provincial distribution centres which had been equipped with primitive printing equipment (Oram, 1983: 262–4). Its drive into the Dublin suburbs was further enhanced by an editorial policy targeting a particular suburb every week, prominently featuring selected news stories from these areas and supporting sales with a poster blitz in the localities concerned. Its well-deserved reputation for hard news received an extraordinary boost when, on three separate occasions in its first five months in existence, it tracked down babies taken from their prams by mentally distressed women, and saw them triumphantly returned to their mothers.

More serious business was afoot at its stablemate, the *Irish Press.* The appointment of novelist Benedict Kiely as its literary editor in 1951 made the paper's weekly literary pages a haven for younger Irish writers, and the paper's politics were also undergoing a discreet change. De Valera's surveillance was becoming more remote, and his deputy Sean Lemass' tenure of office as the paper's managing director in 1948–51 had left its mark. Specifically, the paper was now beginning to take sides in an internal argument within Fianna Fáil. That party's old guard were firmly convinced that the best programme for economic development in Ireland remained one based on native industries, using native raw materials and protected by tariff walls. This policy was running into the sand in the post-war years. The native industries shielded by protection were inefficient, given to taking easy profits as near-monopolies. Imports, because of the tariffs and a licensing system, became the preserve of another coterie of comfortable agencies. Economic growth was stagnating and emigration increasing.

Lemass, who had overseen this policy since the 1930s, now recognized its inadequacy and challenged his party's prevailing orthodoxy in a number of provocative speeches, notably in 1954, in which he chastised the shortcomings of native industries and warned that they could not expect protection to continue indefinitely. This provoked a subterranean ideological dispute within the party, in which the *Irish Press* played a significant role. It did so by subtly endorsing Lemass' side of the argument, highlighting his attacks on the weaknesses of Irish industry, and sidelining or minimising contrasting party views.

The late 1940s and early 1950s were marked by a number of significant changes in radio broadcasting from Dublin (and Cork), based partly on the increase in resources for the doomed short-wave station. At a technical level, there was a vitally important change from disk to tape recording. Financially, there was a considerable improvement in licence-fee revenue after a tough enforcement campaign (about a quarter of listeners had neglected to pay). Restrictions on advertising were considerably relaxed after 1949, and content was changing to admit not only the dreaded sounds of jazz – a new series on this topic was launched in January 1948 – but a whole series of programmes for sports fans, children and undiscovered artistic talent (Gorham, 1967: 183–5).

Radio Éireann remained primarily a public-service organization, structured along sub-Reithian lines, in that it was more closely controlled than the BBC, and primarily used for politically driven ideological ends. The latter factor was evident in relation to the Irish language. The new government that had come into power in 1948 was determined not to be outflanked on a number of issues that its predecessors had made peculiarly their own, notably those of partition and the Irish language. It moved, for example, to increase the time allocated to news broadcasts in Irish, but the way it did so provided as much evidence of inability to tackle the problem as of enthusiasm for the cause.

Part of this inability was systemic. Since independence in 1922, the Irish political and administrative system, though based on a network of local authorities to complement national government, was highly – and increasingly – centralized; local authorities had relatively little control over resources. This militated against a proactive approach to the revival of Irish along the lines of the 1948 subsidy to the Irish-language print media, but also against establishing a radio station dedicated to the needs of Irish speakers living in the Gaeltacht communities. This had been mooted for at least two decades, but had consistently been rejected.

The political bias against decentralization in broadcasting was buttressed by other factors working against the use of radio as a tool to revive the Irish language. One was undoubtedly financial: the department of finance, always apprehensive of expenditure on broadcasting, resisted further 'extravagance' on this modern luxury. Another was ideological: the maintenance of intensive language instruction in the schools perpetuated the notion that the entire population was, at least potentially, Irish-speaking; to encourage the establishment of a separate Irish-language station would have amounted to a damaging admission that this was not the case. This was coupled with the danger that an Irish-language station would allow the English-speaking population to avoid the exposure to the ancestral language intrinsic to a single-channel, bilingual monopoly. The third was geographical: Ireland's Irish-speaking population was scattered, as the language retreated before the advance of commerce and was weakened by emigration, which differentially affected the Irish-speaking areas in the north-west, west and south. There were therefore three strong regional dialects, each with its political and academic defenders, who fought vigorously to establish pre-eminence even as the number of speakers (in any dialect) dwindled inexorably. By 1950, even though the census reports indicated a spread of Irish outside these traditional areas, the number of Irish-speaking monolingual citizens, already small at the turn of the century, was in decline.

This led to the situation in which Radio Éireann had no separate service for Irish-language listeners. Its news in Irish was written in English then translated into Irish. Different dialects were used on successive days to humour the champions of each. It was to be more than another decade before any degree of standardization was achieved.

Other, more political issues were also beginning to surface. From 1932–48, it had been tacitly accepted that Radio Éireann was effectively a government mouthpiece, at least insofar as news and current affairs were concerned. Ritual complaints about this were made from time to time, without any expectation that things would change. Opposition spokesmen were allowed on the radio at certain times, but in a context that spoke of concession rather than of right. There were no live political discussions.

The sensitivity of the situation was exemplified by a controversy that broke surface only after the 1948 change of government. One of the station employees (all of whom were at this time ordinary civil servants), C.E. Kelly, was also the proprietor and editor of *Dublin Opinion*, the humorous monthly journal. His mild satires on political figures were seen by Fianna Fáil as *lèse-majesté* of a high order, and his political motivation was publicly queried in the Dáil. The new government retaliated by making him director of broadcasting when a vacancy arose. Although he was even-handed throughout, Fianna Fáil redeployed him rapidly away from direct involvement in broadcasting after its return to power in 1951.

The 1948–51 period, although marked by financial and technological advances, was hardly notable for innovation in broadcasting. Indeed, with the exception of an unprecedented excursion to Rome in September 1950 to record the Holy Year celebrations – the opposite was the case. Plans for schools broadcasting were aborted, and the advice of the Broadcasting Advisory Committee, which recommended in 1950 that broadcasting should be removed from direct government control, was ignored. The change of government in 1951, however, saw the appointment of a new minister, Erskine Childers, whose vision for the station, adumbrated in some detail in a major speech in the Dáil in November 1952, gave a completely new tone and impetus to the direction, content and control of Irish broadcasting.

Childers was a moderniser, but he was also given to direct interference in broadcasting. He was among the few Irish politicians of his era who saw broadcasting as an opportunity rather than a threat, and his structural changes, although in many respects tentative, were a sign of things to come. He effected a good working relationship with the new secretary of the department of posts and telegraphs, Leon Ó Broin, from whom, unusually in the case of a civil servant, many of the decentralizing ideas had emanated. Childers took ownership of these ideas, although his enthusiasms have to be put in the context of his sometimes overweening interest in minutiae: broadcasting or administrative staff would on occasion receive memos from him containing up to twenty-two separate queries, each requiring a specific answer (Gorham, 1967: 222).

His structural changes mirrored his character as well as his objectives. He proposed the establishment of a *comhairle* (council) of five people to advise the minister, and be responsible under him for the general control and supervision of Radio Éireann. The downstream effects of this had to be worked out in practice: this included defining the relationship between this new council and the old Broadcasting Advisory Council, which the minister had a statutory obligation to maintain, and the financial and staffing relationships between the department of finance, the department of posts and telegraphs, and the reorganized broadcasting service. One thing was clear: the new *comhairle* would have only as much power and responsibility as the minister permitted, and the minister retained overall control, including direct control if necessary.

Although the minister now had two councils, the old one and the new one, the relationship between them was never satisfactorily clarified. The minister intended to constitute the old council as a listeners' representative body, and the new one as a management body; but the possible flaws in this model led to a decision not to reappoint the old council when its mandate expired in November 1964, despite the statutory requirement to do so.

Childers' changes to staffing and financial arrangements were substantial by contemporary standards. Large numbers of staff were transferred from the department of posts and telegraphs to work under the new council. Some made the transition from being traditional civil servants to administrators of a broadcasting system with more difficulty than others. But the new arrangement also benefited from a much greater degree of financial autonomy, and clarity in relation to broadcasting budgets, than had existed heretofore under the regime of the department of finance.

Childers showed his personal style in two areas in particular. One was in his encouragement of advertising and sponsored programmes. Before becoming a member of parliament in 1944 he had worked as advertising manager of the *Irish Press* and as secretary to the Federation of Irish

Manufacturers, and he had a long-standing acquaintance with many industrial and commercial concerns. In March 1952, Irish companies that imported goods for sale in the Republic were allowed to advertise on radio for the first time, as long as their products were not in competition with any Irish-made articles. Broadcast finances were further improved by an increase in the licence fee, although the costs of collection remained inordinately high, and there was a high level of evasion.

The second related to content. Like P.J. Little, Childers' enthusiasm for music was evident in his continuing support for the station's orchestral output. Other cultural innovations included the inauguration in September 1953 of a public lecture series, the Thomas Davis Lectures (named after the nineteenth-century Protestant nationalist leader), more programming for children and from the regions, and the first broadcasts of news in the mornings (in the middle of a newspaper strike in July 1952). In the absence of an effective short-wave service, an unexpected substitute was found: Brazzaville, in the French Congo, was prepared to rebroadcast Irish programmes, particularly sports reports, which were avidly listened to by a very select sub-group of the Irish diaspora – Irish missionaries working in Africa. It was a long way from Eamon de Valera's project of a primitive World Wide Web for Irish emigrants everywhere on the globe, but it at least met an identifiable, if limited, need with rare precision.

Political broadcasting, however, was the area on which most new ground was being broken. Under Childers, permission was given for the first time for unscripted political discussions; the ban on members of parliament participating in broadcasts was removed; and, in the run-up to the 1954 general election, party political broadcasts were introduced for the first time. The political parties reacted to some of these initiatives with suspicion, and the larger parties discovered that, by refusing to participate in political-discussion programmes, they could stymie discussion on any topic on which they felt uncomfortable – and did so with frequent effect.

Under Childers' new council, broadcasting developed an empirical basis, inaugurating listener research via surveys in 1953, 1954 and 1955. The first enquiries revealed that 85 per cent of the potential audience listened to Radio Éireann, compared with 53 per cent for Radio Luxembourg and 49 per cent for the BBC Light Programme (Gorham, 1967: 229), and that there was an unexpectedly large audience for Irish dance music programmes. This was an oddity given the visual nature of Irish dancing, but the popularity owed much to the personality of one its best-known presenters, known as 'Din Joe'.

Audiences for programmes in the Irish language rarely achieved double figures, but this was perhaps to be expected. As Maurice Gorham noted mildly, Radio Éireann was:

> expected to revive the speaking of Irish; to foster a taste for classical music; to revive Irish traditional music; to keep people on the farms, to sell goods and services of all kinds, from sausages to sweep tickets; to provide a living and a career for writers and musicians; to reunite the Irish people at home with those overseas; to end Partition. All this in addition to broadcasting's normal duty to inform, educate and entertain. And all in a programme time amounting (if advertising time was excluded) to some five and a half hours a day. (Gorham, 1967: 222)

As the listenership figures for Radio Luxembourg and the BBC plainly indicated, Radio Éireann did not have the luxury of meeting these objectives in a vacuum. Its potential radio rivals, however, were now to be crowded almost off the stage by an even more potent competitor: television. In 1950, the secretary of the department of posts and telegraphs, Leon Ó Broin, had inaugurated a long war of attrition to convince his colleagues in the department of finance to take television seriously. Initially, suspecting costly empire-building on the part of Ó Broin, the department of finance rebuffed his attempts brusquely. In April 1951, a commercial manufacturer of television sets, the Pye company, organized a demonstration of television at the Royal Dublin Society, which created immense public interest. Despite finance's disapproval, Ó Broin inaugurated small-scale explorations of the topic.

By 1953, BBC television from Wales and Yorkshire was available in parts of the Irish eastern seaboard. When the BBC opened its first transmitter in Belfast late in that year, the social and political reverberations were felt in Dublin (where viewing was free, as no television-licence scheme was in existence). Ó Broin had set up a small departmental committee to study the problem, but Childers did not readily warm to the innovation. Too much television, he warned the Dáil, would kill the art of conversation, and he would never recommend the establishment of any Irish television service 'that is not absolutely first class and directed towards the preservation of the national culture as well as for entertainment purposes' (DD, 10 Nov. 1953). Positions were being prepared for the lengthy and complicated ideological and financial arguments that were to frame the eventual introduction of an Irish television service eight years later.

In the Northern Ireland print media, the early post-war years changed little in the *Irish News* or the *Belfast News Letter.* Technological developments were few or non-existent, and staffs remained small: the entire editorial staff of the *Irish News* during this period amounted to less than a dozen journalists (Phoenix, 1995: 34). The principal event at the end of this period was the IRA campaign against Northern Ireland police barracks from 1956 onward, which exercised the attention of Northern Ireland newspapers to a predictable degree thereafter.

The context was broadly political in that in 1949 the Southern government had launched an organization called the Anti-Partition League, an all-party body designed to remove the partition issue from the sphere of domestic politics. This organization became more proactive than Southern politicians had been, establishing branches overseas and subventing political activity by nationalists within Northern Ireland. It also had a journalistic context. The *Sunday Press*, founded in 1949, regularly published features extolling the lives and exploits of the IRA during the War of Independence. Noel Browne accused de Valera in his autobiography of 'shamelessly fostering the cult of the warrior and the soldier' in his papers during this period (Browne, 1986: 232). Certainly the governments between 1948 and 1956 had put themselves in a cleft stick by increasing the ideological and political pressure for the removal of the border, while decrying the paramilitary expression of entirely similar sentiments. There was an equal dilemma for nationalist newspapers such as the *Irish News*, which was to some extent alleviated by the prompt action of the Catholic bishops in condemning the new IRA campaign. The *News* was as even-handed as any Catholic and nationalist newspaper could be in the circumstances, declaring in an editorial on 27 December 1956 that 'much unhappiness in Ireland springs from the British-made border ... But the unhappiness will be increased by acts that defy the instruction of the bishops and weaken Ireland's position as a Catholic nation' (Phoenix, 1995: 33).

The *Belfast Telegraph* was undergoing something of a transformation. Its managing director, Bobbie Baird, an enthusiastic amateur racing driver, died in a racetrack accident in July 1953, necessitating a reshuffle of responsibilities (and touching off a series of financial problems that was to end, nine years later, with the purchase of the paper by the Thomson organization). The editor, Robert Sayers, was appointed chairman as Baird's successor and was succeeded as editor by his nephew Jack Sayers, whose own father had edited the paper in the 1930s. At one level a merely dynastic change, it also heralded a somewhat unexpected change of style and tone. So far impeccably unionist, the *Telegraph* now became considerably more open to other agendas. John Cole, who worked on the *Telegraph* as a young journalist under Jack Sayers and was later to rise to eminence in London journalism, summed him up:

> Sayers wanted to make the paper more politically serious and intelligent, a purpose with which I strongly agreed, though I was sometimes out of line on some issues. He was determined to influence politics in Northern Ireland. He had returned from the navy to be appalled by the narrowness of political attitudes. He was a liberal unionist, and wanted to build bridges with the nationalist community. His editorial policy was eventually to alienate some influential Unionists, businessmen and others, and various management figures, I gather, made alarmed noises about the dangers of losing advertising, but Jack held firm, knowing that the advertisers needed the *Telegraph* as much as it needed them. (Brodie, 1995: 105–6)

By the time of the Stormont general election of 1958, Sayers was emboldened to open the feature pages of the newspaper to opposition parties. He welcomed the election of four Labour MPs on that occasion as a sign of the coming normalization of Northern Ireland politics, and, in his post-election editorial of 1 April, noted:

> The Unionist Party must re-emphasise that it is a party of the centre. There is being created a body of moderates, Catholics among them, whose votes are being mobilised and will go, outside the most industrial constituencies, to the party that offers reasonable discussion and impartial dealing, true civil and religious liberty and equality of citizenship.

Official unionism was not impressed, but the gamble – if indeed it was a gamble – paid off handsomely. Profits increased, allowing the purchase of state-of-the-art printing machinery in 1955, and by 1957 average daily sales were reaching between 195,000 and 200,000 copies, an extraordinary performance that made the *Telegraph* the largest-selling daily newspaper on the island. Sayers continued to preach his brand of liberal unionism – even resigning in protest from the Orange Order in 1959, at a time when unionism had officially set its face against even the possibility that Catholics might be allowed to join the Unionist Party (Gailey, 1995: 57–8). But his paper missed one vital trick, forgoing a potentially lucrative investment in television.

The arrival of television was to reopen a number of old sores, and create a few new ones. The debate had begun as early as 1951, with the publication of the Beveridge Report on new structures for the BBC in the era of television, but the proposal for each region to have its own commission, broadly representative of its community, was so bedevilled by the divisions in Northern Ireland society that such a model was seen as inoperable there. A broadcasting service designed for the whole community, the *Northern Whig* pointed out on 12 June 1952:

> should preserve a greater aloofness from possibly controversial matters, should not only be aloof but should be known to be aloof, and though in the result it may be less representative of the province than might otherwise be wished, the gain must outweigh the loss. (Cathcart, 1984: 167)

BBC television arrived in Northern Ireland in the spring of 1953, but the coverage of the booster station was initially small. It was not until a new transmitter on Divis Mountain outside Belfast came into operation in July 1955 that coverage reached some 80 per cent of the Northern Ireland population. The first live broadcast from the Belfast studio – including an interview with the Northern Ireland prime minister, Lord Brookeborough – took place in November 1955. This was also the year in which the first live television broadcast – an Ireland-vs-England boxing tournament – took place on the other side of the border. This was also organized by the BBC and was relayed via Belfast and Scotland to a number of other European countries (Fisher, 1978: 23).

When the introduction of commercial television was mooted in 1957, the *Belfast Telegraph* negotiated with a number of newspaper and other interests to join a consortium bidding for the Northern Ireland Independent Television franchise. The board, however, was divided, and in December 1958 opted out of the consortium. The franchise eventually went to another consortium, headed by the *Belfast News Letter* (Brodie, 1995: 119–20). It was to prove an expensive decision.

The arrival of Ulster Television (UTV) – the Northern Ireland franchise-holder in the Independent Television network – at Halloween 1959 was an important staging post in a series of developments initiated almost a decade earlier (Cathcart, 1984: 156–89). In 1953, moving into a slightly higher gear, BBC Northern Ireland invited Professor Theo Moody, a Quaker, who was professor of Irish history at Trinity College Dublin, to be one of the organizers of a series of historical lectures broadcast (and subsequently published) under the title *Ulster Since 1800*. This series, while avoiding the potentially explosive recent past, constituted a serious attempt to tackle contentious subjects. Moody, according to Cathcart (1984: 176), used this experience to persuade the Dublin broadcasting authorities to inaugurate their regular Thomas Davis Lecture Series that same year.

In January 1954, an unscripted discussion programme, *Your Questions*, showed that people could disagree politically on the airwaves in a civilized manner. In the same year, just after Radio Éireann, BBC Northern Ireland began scientific audience research, which revealed that – in sharp contrast with the other BBC regions – local programming was more popular than the network output (Cathcart, 1984: 182). In 1956, a BBC radio studio in Derry was blown up by the IRA at the beginning of their armed campaign against Northern Ireland institutions. The explosion, like other aspects of that campaign, had no discernible result on broadcasting policy, although it contributed substantially to understandable unionist defensiveness in general.

The arrival of BBC television in Northern Ireland, and the fact that it was followed in 1959 by UTV, not only created a new set of tensions within Northern Ireland itself, but had a demonstrable effect on the other side of the border, where the secretary of the department of posts and telegraphs, Leon Ó Broin, was trying to persuade the government to adopt a realistic attitude to the new medium. Political competition between the two parts of the island was to give his campaign the fillip it desperately needed.

5 Broadcasting, 1957–73

The late 1950s gave rise to critical debates about the nature, control and function of television in both parts of Ireland. Initially, and especially in the South, the debates were primarily about structures and control. In the late 1960s, questions of content began to assume greater significance, partly because of the developing Northern Ireland political crisis, partly because of social – and other – changes ruffling old certainties.

In the Republic, a posts and telegraphs committee report presented to the government in September 1953 pushed for the establishment of a limited television service confined initially to Dublin, on the grounds that the republic could not be seen to be lagging behind Northern Ireland. This political agenda was buttressed by the suggestion that a wholly private, profit-driven service would undermine the protection and development of traditional Irish culture. Given that the new service would be a monopoly, the committee argued that it should be under state control; given that it would be expensive, it should be financed by advertisements as well as by a licence fee; and, given that direct governmental control was inappropriate, it should be run by a government-appointed body (Savage, 1996: 21–7).

Initial arguments about this report were coloured by finance's misapprehension about what Leon Ó Broin, secretary of the department of posts and telegraphs, was proposing. They assumed he envisaged a television service run by the department of posts and telegraphs, but Ó Broin believed in giving broadcasting a measure of autonomy. All these arguments were, however, to some extent advanced in a vacuum, both because of finance's opposition, and also because of a generally negative attitude towards the electronic media – Dublin's politicians tended to mirror the belief of their Belfast counterparts that broadcasting was a necessary evil.

By early 1954, Childers had had a change of heart. His initial hesitancy overcome, he began to push for the development of television, or at least for a planning process to begin. Before his enthusiasm could gather momentum, however, his government lost the 1954 general election, and were succeeded by a second inter-party or coalition government, whose minister for posts and telegraphs, Michael Keyes, was cool towards the idea. Pressure, however, was building up, especially after the enhancement of the BBC signal strength in Northern Ireland in 1955, and the reception of the first UTV signals from mainland Britain. Politicians and journalists alike began to question the government's lethargy. An editorial in *Comhar* asked in September 1953: 'Can we afford … to do nothing to minimise this baneful influence? Are we satisfied to let England win the last fight so easily?'

Private companies were also beginning to nibble at the edges of what appeared, to some of them at least, a potentially profitable proposition. Politicians were drawn into the action: a senior member of Fine Gael, Sean Mac Eoin, acted as a lobbyist for one group; Childers himself, now in opposition, was arguing enthusiastically for the interests of a competitor, the UK firm Pye Ltd.

All these developments prompted Ó Broin and his committee to submit a further report to government in March 1956. A new argument was added to those already adduced in favour of a native television service: the danger to Irish morals emanating from foreign broadcasts. The BBC's television output was governed by concepts 'wholly alien to the ordinary Irish home'. Some offending programmes were:

> brazen, some 'frank' in sex matters, some merely inspired by the desire to exalt the British royal family and the British way of life … [T]here is constant emphasis on … the British view of world affairs, and the British (including Six County) achievements. The BBC's resources are lavishly expended on reporting every movement of the royal family at home and on tour. (Savage, 1996: 46)

Robert Savage's seminal 1996 study of the origins of Irish television suggests that this argument was a natural expression of the sexual conservatism and political isolationism characteristic of 1950s Ireland. However, this presupposes an unusual degree of personal prudery on the part of Ó Broin and his committee. It should be remembered that Ó Broin was an exceptionally intelligent and literate civil servant (he wrote excellent history in his spare time), who, like most civil servants of his generation, was skilled in identifying and emphasizing those arguments that might best appeal to ministers. The cabinets to which he addressed these reports were, for the most part, composed of ageing men close to retirement. In addition, the 1954–7 coalition cabinet was notably more clericalist than its Fianna Fáil predecessor, and might have been expected to pay serious attention to the warning of Pope Pius XII in 1955 that:

> It is impossible not be horrified at the thought that through the medium of television it may be possible for that atmosphere poisoned by materi-

alism, fatuity and hedonism, which is too often breathed in many cinemas, to penetrate within the very walls of the home. (Savage, 1996: 110)

On the other hand, there was a central fallacy in this argument, of which Ó Broin and the committee could hardly have been unaware, and which supports the possibility that their emphasis on the moral issue may have been more tactical than ideological. British television broadcasts were already, by the late 1950s, available to about 40 per cent of the population of the Republic. To suggest that the advent of an Irish television service would eliminate or even dramatically reduce the attractiveness of British television, with its huge budgets and sophisticated production values, would have been extraordinarily unrealistic. And broadcasts featuring the British royal family were even more problematic. Irish audiences have long combined a deeply rooted and at times even visceral republicanism with a deep fascination for the house of Windsor. During the 1981 wedding of Prince Charles and Diana Spencer, social activity outside Irish workplaces came to almost a complete halt for a period of several hours. It is likely that Ó Broin and his colleagues knew that, even if they could not turn back the tide, they could harness some of its hydro-electric power to their own purposes.

Ó Broin was prepared to find allies anywhere – even in the BBC, whose programmes he had sometimes distrusted. Realising that the cost of a new Irish television service could be sharply reduced if some BBC programming – evidently not the sort with which he found fault – could be acquired at discounted rates, he tried unsuccessfully to get a commitment from the BBC in this regard. Although sympathetic to his argument that such an arrangement would help to create a barrier against broadcast material imported from the US (which had provisionally offered to give Ireland an advertising-funded television service at no cost), the BBC rejected this approach and so, in due course, did the government (Horgan, 1997: 319).

After the change of government in March 1957 the cabinet began to discuss the problem in an earnest, but apparently confused way. It decided in October to establish a television service 'as early as practicable ... under public control ... [and] so far as possible, without cost to the Exchequer' (NAI, Taoiseach: S 1499B). Less than a fortnight later the new minister for posts and telegraphs, Neil Blaney, made a somewhat ambiguous announcement in which he intimated that the new service would be largely commercial in character, and that its entire capital and maintenance costs would be provided by a promoting group or groups licensed to operate a service of commercial programmes for a term of years (*IT*, 7 Nov. 1957). It would, he added, be a condition of the licence that a certain proportion of the schedule would be available for programmes of a public-service character.

This announcement touched off a major controversy. Senior broadcasting executives opposed to further commercialization threatened to resign, an (unrelated) cabinet reshuffle moved Blaney elsewhere, and Ó Broin redoubled his efforts. His position received significant support from a senior political figure, the minister for finance, Sean MacEntee, who secured his colleagues' agreement early in 1958 for the establishment of a commission to look at the whole question, but whose terms of reference were expressly written to ensure:

> that no charge shall fall on the exchequer, either on capital or on current account, and that effective control of television programmes must be exercisable by an Irish public authority to be established as a television authority. (Keogh, 1995: 33)

The dilemma was clear. The only way to ensure that television would be cost-free appeared to be to entrust it to a commercial organization; but no commercial organization would be prepared to take it on board with the restrictions envisaged. Initially, this did not dissuade would-be suitors, including Pye and the Canadian media magnate Roy Thomson. Gael Linn, the Irish cultural and language organization, also offered to run the new service. After the publication of the commission's report, but before the government had taken its decision, there was also a late and indirect approach from the Vatican, where the pope had apparently overcome his dislike of television to the extent that he was prepared to accept that, in safe, Catholic and Irish hands it could be used as a medium for the re-evangelization of continental Europe (Savage, 1996: 154–5).

The difficulties encountered by the commission in reconciling its brief with the political and economic realities of the situation were evident in its final report, submitted to the government in May 1959. Four out of twenty members submitted a minority report, and even some who signed the main report expressed personal reservations about aspects of it.

The commission concluded that handing over television to a private monopoly was preferable to having it rely on public subsidy – given the way in which its terms of reference had been drafted – the only way in which no charge would fall on the exchequer. It advised the government that:

if Ireland is to have a television service, and under the existing circumstances the commission does not accept that Ireland can afford to be
without its own television service, it follows that the service must for the
present be provided by private enterprise, notwithstanding the considerable difficulties that are attendant on the establishment of a
commercial service by private enterprise. (Television Commission, 1959:
23)

It also – and this is where its attempt to square the circle became most
obvious – suggested that the public authority envisaged in its terms of
reference should be appointed by, and answerable to, the government, and
should have important programming functions. And it recommended that
radio and television should be under two separate public authorities.

The completion and submission of the commission's report evoked a last,
desperate intervention from Ó Broin. In a memorandum to the government,
written ostensibly on behalf of his minister as a commentary on an interim
report from another government commission, set up to recommend ways of
protecting the Irish language, he combined criticism of some of that
commission's more unrealistic suggestions with a subtle appeal to the known
cultural proclivities of the cabinet. A programme contractor of the type
envisaged, he argued, would be solely motivated by profit, and would ignore
national objectives – including the revival of the Irish language:

If television is to be used positively for national objectives, the proper
procedure in the Minister's view is that the Television Authority should
itself operate the programmes – that is – all the programmes. (NAI,
Communications: T007/57)

Two months later, the taoiseach, Eamon de Valera, resigned, and was
succeeded by Sean Lemass, who favoured the privatization option.
Remarkably, one of the first cabinet meetings over which Lemass presided, on
31 July 1959, rejected the commission's core proposal, and decided instead that
the new service should be provided under a public statutory authority alone.

On the face of it, this was an extraordinary volte-face by Lemass, but there
is strong circumstantial evidence that he was an unwilling convert. In the first
place, he tried to keep the door open for one of the unsuccessful foreign
bidders even after the decision had been taken, attempting to persuade his
minister for posts and telegraphs to interest the disappointed businessman in

establishing 'profitable' international radio broadcasting from Ireland (NAI, Communications: T007/57). Secondly, after his retirement he leaked the fact that his cabinet colleagues had out-voted him on the issue (*IP*, 18 Jan. 1969). De Valera had departed but Lemass had inherited a cabinet that retained a strong majority favouring the older man's point of view. They would also have been far more receptive than Lemass to Ó Broin's final, successful plea (Horgan, 1997: 312).

The broadcasting legislation drafted on foot of this government decision became law as the 1960 Broadcasting Act. By now, the number of radio licences had increased to some 500,000. The number of television sets, though increasing rapidly, was still probably less than 50,000. The act set up a new authority, known as the Radio Éireann Authority (it later became the Raidió Teilifís Éireann (RTÉ) Authority), whose members were appointed by the government. In general terms, the authority was given a relatively free hand in managing the new service, but the government retained a number of functions. These included determining the length of the broadcasting day, and the amount of advertising permitted. Section 17 of the act required the authority to 'bear constantly in mind the national aims of restoring the Irish language and preserving and developing the national culture'. Section 18 required that – in news, current affairs and controversial subjects – the station present material objectively and impartially, and without any expression of the authority's views.

The only other section of the act dealing with programming, which would later prove one of its most controversial aspects, was Section 31, which empowered the government to issue a directive to the authority prohibiting the broadcast of specific material, or insisting on the broadcast of specific material. Although the IRA campaign against the British presence in Northern Ireland that had started in 1956 continued, it was of a very low intensity and not mentioned in this debate about broadcasting. It was to be a decade before Section 31 was first implemented, at the height of the conflict that was to erupt in Northern Ireland in 1969.

The 1960 act was, according to one authority, a relatively forward-looking piece of legislation, given the era in which it was drafted:

> It gave the broadcasting authority a great measure of autonomy, protected by legislation, and it ensured that any government directives which would affect programming had to be served on the Authority in writing. Given the particular circumstances of Ireland, still a compara-

tively young state, born in violence and immediately afterwards torn by civil war, with a continuing internal security threat, an unsolved problem as regards Northern Ireland, economically under-developed and socio-logically unsettled, it was a remarkably liberal piece of legislation and, by and large, provided a sound statutory framework for the first fifteen years of the re-structured broadcasting service as it entered the television age. (Fisher, 1978: 27)

The context within which the act was introduced, however, was more volatile than either government or broadcasters knew. De Valera's resignation in 1959, after more than thirty years as leader of his party – twenty-one of them as taoiseach – marked the departure of the political old guard. Leadership in the other major political parties was changing too. The economy was improving, with the introduction of Keynesian economic planning in 1958; the level of popular education was rising; the emigration that had drained the country of its most vital human resources, especially in the 1950s, was lessening; urbanization was increasing; and – whether prompted by the pernicious television programmes identified a few years earlier or not – social attitudes were undergoing an accelerated process of challenge and change.

Lemass, although himself a moderniser, was conscious of the dangers inherent in the new medium, and at one stage, even before the new service had been inaugurated, contemplated using Section 31 to order the authority to pay particular attention to the 'image' of Ireland presented by television, 'including the avoidance of stage-Irishisms, playboyisms etc'. Insofar as social problems were concerned, he thought the desirable course would be to encourage objective presentation of facts and constructive comment. 'The "God-help-us" approach should be ruled out', he instructed. He had origi-nally planned to codify these instructions as part of a formal government directive to the new station, but was dissuaded from this course of action by a senior civil servant who warned him this approach might appear illiberal (Keogh, 1995: 32).

The inauguration of the new service, on New Year's Eve 1961, was notable for a number of things. One was the doom-laden tone of de Valera, now president of Ireland, as he inaugurated its first broadcast: 'I must admit that sometimes when I think of television and radio and their immense power I feel somewhat afraid ... [it] can lead through demoralisation to decadence and dissolution' (*IP*, 1 Jan. 1962). Another was the broad welcome from the print media, apparently as yet unaware of the threat TV would pose to their advertising revenues. Critics wrote a daily review of the programming, and, although some of it was criticized for having 'very little protein content' (*II*, 2 Jan. 1962), there was a general air of excitement and celebration.

Television soon became a battleground in a way radio had never been. Some of the critics gave the appearance of having been lying in wait in the long grass: Revd P. O'Higgins, SJ, assistant director of the Pioneer Total Abstinence Association, quickly complained that he was 'appalled to see on one of the commercials very young people making whoopee with drinks in their hands' (*II*, 18 Jan. 1962). Subsequent conflicts in the areas of politics, culture and religion were more intense.

By simultaneously setting up a new service – television – and removing all broadcasting from the direct control of government, the cabinet had effectively taken two large steps in one. They retained a residual right of control under Section 31, and the right to dismiss the authority, but these were guns that could each be fired only once. There were no formal channels through which they could communicate their views to the authority, short of issuing a directive, and if they were dissatisfied with or annoyed by the station's output they were required (short of dismissing the authority) to formally suffer in silence.

Silence does not come naturally to the average cabinet minister, and early forays by Irish television commentators and presenters into politics and current affairs made many of them bristle. The tentative but – relatively speaking – venturesome initiatives of some broadcasters were fuelled by a desire to hold politicians to account in a way media had rarely done before. This was accentuated by the weakness of orthodox political opposition: the only alternative to a Fianna Fáil government was a coalition, but the Labour Party had set its face firmly against going into government with any other party. Under these circumstances, the eager young graduates in politics and history who filled the new current-affairs programme slots often found themselves making an opposition case by default. Government ministers, unused to the journalistic conventions whereby interviewers ask provocative questions that do not necessarily reflect their own point of view, sometimes mistook liveliness for wholesale subversion. The RTÉ current-affairs programme *Broadsheet* was regarded as a prime offender. Lemass, as taoiseach, frequently intervened privately with the director general of the station in attempts to secure a fairer hearing for government policy on this particular programme (Horgan, 1997: 316). It was, he told an aide in 1962, becoming increasingly 'a medium for the uncritical presentation of the views of persons associated with various ramps and crank projects', and broadcasters should 'take the whine out of their voice' (NAI, Taoiseach: S 3532 C/63). In 1963, his son-in-law, future taoiseach C.J. Haughey, took up the cudgels in a spirited but unsuccessful attempt to ensure that when ministers or government departments issued statements of special importance, 'the statement … should be reported verbatim or not at all' (*IT*, 1 Jan. 2002).

The most dramatic conflict in this area took place in 1966, shortly before Lemass retired. The government was at odds with the National Farmers' Association over agricultural policy, and it emerged that the relevant minister had contacted the television station unofficially to complain about a lack of balance in a news bulletin on the dispute. Queried about this in the Dáil, Lemass made no bones about it: the government, he said, would make whatever representations might be necessary to ensure that Raidió Teilifís Éireann did not deviate from the performance of its duty:

> Radio Telefís Éireann was set up by legislation as an instrument of public policy and as such is responsible to the government. The government have over-all responsibility for its conduct and especially the obligation to ensure that its programmes do not offend against the public interest or conflict with national policy as defined in legislation. To this extent the government reject the view that RTÉ should be, either generally or in regard to its current affairs and news programmes, completely independent of government supervision. As a public institution supported by public funds and operating under statute, it has the duty, while maintaining impartiality between political parties, to present programmes which inform the public regarding current affairs, to sustain public respect for the institutions of government and, where appropriate, to assist public understanding of the policies enshrined in legislation enacted by the Oireachtas. (DD, 12 Oct. 1966)

This gloss on the Broadcasting Act provoked widespread controversy. Although it might be going too far to suggest that Lemass was equating 'public policy' with 'government policy' (Doolan, Dowling and Quinn, 1969: 205), it was unmistakably the first public exchange in a hitherto subterranean conflict. RTÉ's broadcasters retaliated spiritedly, with a week of programmes dedicated to the issue of media freedom, and even managed to internationalize the controversy by securing an airing for their views on US television. This failed to produce any very visible change in government practice. In 1967, the government intervened directly to prevent the station sending a television crew to North Vietnam (NAI, Taoiseach: 98/6/19), and, in January 1968, to prevent another crew being sent to the breakaway state of Biafra, which was attempting to secede from the Federal Republic of Nigeria, and in which many Irish Catholic missionaries were working. The dangers of a clash between the broadcasting agenda and Irish foreign policy, which sought to avoid unnecessarily offending the legal Nigerian government (a military dictatorship), and which favoured the US role in world affairs, was the prime motivating factor in each case. In 1969 there was another major controversy involving a current-affairs programme on illegal moneylending in Dublin. The Garda Síochána were outraged by the programme's suggestion that they were not doing enough to stamp out this activity, and they pressured the government into establishing a lengthy judicial tribunal to enquire, not into illegal moneylending, but into the bona fides of the programme itself. Unsurprisingly, this tribunal came to the conclusion that the programme had not provided legally convincing evidence for its allegations.

There were further controversies about the role of current-affairs broadcasting on political affairs, involving lengthy negotiations between party whips and producers about the ratio of participants from government parties to those from the opposition. These skirmishes resulted in an incremental improvement in the freedom of the station and in the willingness of politicians to accept a measure of accountability. The basic mood has been accurately described as one in which current-affairs broadcasters 'have generally aligned themselves with a middle ground which is progressive and social democratic' (Kelly, 1984: 98).

In the area of culture, the Irish language was at the centre of controversy. The authority was criticized from both ends of the spectrum. Irish-language organizations took up the running initially, claiming that the authority was failing to fulfil its obligations under the act because it did not broadcast enough programming in Irish. Later, and especially after the foundation of the Language Freedom Movement in 1964, dedicated to abolishing compulsory Irish in schools and as a requirement for positions in the public service, it found itself criticized for broadcasting too much. (Television remained a primarily urban phenomenon: a 1967 survey found that the proportion of rural households with television was as low as 25 per cent in some areas; in 1978, only 54 per cent of households in Connacht, the region with the greatest number of native Irish-language speakers, had televisions, compared with 92 per cent in Dublin (Brown, 1981: 261).)

The unspoken argument was that RTÉ had to pay its way, and that leaning too far in the direction of a proactive policy on Irish might encourage viewers to look more to UK channels, undermining RTÉ's position in the market-place. In the circumstances, it had to steer an unsteady middle course, which ended up satisfying no one. It rejected the idea of ghettoizing Irish-language programming, devoted substantial resources to children's programmes in Irish, and contributed to the popularization of Irish traditional music that was to achieve an international dimension in the 1970s and 1980s. After the publication of the report of the Commission on the Restoration on the Irish Language in 1964, the pressure on RTÉ to play a larger role in the language-revival movement became so strong that it drove the authority's chairman, the broadcaster Eamonn Andrews, to resign (Horgan, 1997: 320).

Pressure continued to mount, and the formation in 1969 of an organi-zation to campaign for greater rights for Irish speakers, Gluaiseacht ar Son Cearta Sibhíalta na Gaeltachta ('Movement on Behalf of Civil Rights for the Irish-Speaking Areas'), intensified protests against RTÉ's perceived inade-quacies in this area. A pirate Irish-language radio station, Saor Raidió Chonamara ('Connemara Free Radio'), was established on 28 March 1970 (Ó Glaisne, 1982: 29), but was quickly snuffed out by the police. Political pressure followed, with questions in parliament, and RTÉ then moved swiftly in 1970 to obviate any challenge to its monopoly by agreeing to establish a separate Gaeltacht station within the existing RTÉ structures. This new station, Raidió na Gaeltachta, began broadcasting in April 1972 from purpose-built studios in Casla in Connemara, and its function was originally envisaged as providing a service from a number of subsidiary studios, purely for the Gaeltacht communities, dispersed across the non-contiguous areas of Connemara, Donegal, Munster and Meath. The first interlinked broadcast from these studios took place in 1973. Over the years, the service gained a listenership outside the Gaeltacht areas: a 1994 survey suggested that it had succeeded in attracting some 15 per cent of the national radio audience, on a regular or occasional basis, during the preceding decade (Watson, 1997: 218).

The question of religion was another, almost equally sensitive area. In 1962, three of the country's Catholic bishops took the unusual step of seeking a private meeting with Lemass in which they complained about the religious views (or lack thereof) of three named broadcasters (Horgan, 1997: 317–18). Lemass defused the situation, and the station moved towards appointing an adviser on religious programmes. Here the bishops suffered a setback.

Archbishop John Charles McQuaid of Dublin had trained one of his younger priests in television and film techniques, confident that this man would be appointed to the new position. The authority, however, looked elsewhere, and employed instead a genial Dominican priest, who, as a member of a religious order, was not subject directly to the archbishop's authority, and who played a delicate and important role during the formative years of Irish television with great tact and good humour. The archbishop's loss, as it turned out, was broadcasting's gain. Dr McQuaid established an independent studio for the young priest, Fr Joe Dunn, who, over the next three decades, with a team of clerical and lay people he recruited and trained, created a series of some 400 television documentaries under the Radharc ('View') label. Many of these were broadcast on RTÉ and also on foreign stations. They represent a remarkable contribution to the genre, and to the social and media history of modern Ireland. The decision to keep RTÉ, and Irish broadcasting generally, at some distance from the power of the Catholic church was further exemplified in 1965 when Lemass responded negatively to a proposal (made to him by Pope Paul VI during an audience in Rome) that Ireland should establish a Catholic radio station capable of transmitting programmes to Britain and further afield. This refusal, however, was somewhat qualified by the minister for posts and telegraphs, Michael Hilliard. 'I have always felt', he wrote to Lemass on 9 April 1965, 'that the proper answer to suggestions for a Catholic station is that we have one already in Radio Éireann' (NAI, DFA: 96/2/14).

It was outside the area of direct religious broadcasting, however, that tensions were sharpest. The advent of a new entertainment programme in the summer of 1962 – it was originally designed as a 'filler' programme, to be taken off again after the summer was over – heralded a new talent, and a new approach that was to have a profound effect on Irish social mores. This was the *Late Late Show*, moderated by Gay Byrne, who had left a career in insurance for one in light entertainment. Byrne's deft choice of topics, and even defter handling of a wide range of personalities (many of them controversial, some of them wild, others dedicated to the art of public confession) immediately set a stamp on the new station that was peculiarly its own. Broadcast on Saturday evenings, the programme became – for an Irish public unaccustomed to the public flouting of taboos – an unmissable weekly occasion. One of the biggest taboos, as might be expected, related to the public discussion of sexual matters, and here the tension was often not just social, but intergenerational, as novelist Colm Tóibín has recalled:

> In Enniscorthy when I was a lad we all sat glued to it. We were often embarrassed that someone was talking about sex: there were older people in the room who didn't like sex being talked about … If any other programme talked about sex, it would have been turned off. Turn that rubbish off. But nobody ever turned the Late Late Show off. The show was too unpredictable: you just never knew what you might miss. (Tóibín, 1990: 87)

But the programme was about more than sex. It was also about politics. Politicians initially rejected its attempts to inveigle them into going before the cameras in such a highly unstructured, and therefore risky, format, and continued to do so until the early 1970s (NAI, Taoiseach: 98/6/20); but they could not resist the lure of its huge exposure to the electorate indefinitely. And it was about religion: its reputation, never seriously in doubt after the early months, grew exponentially when, in 1966, the first bishop preached a sermon critical of the programme. It was also, incrementally, about the development of a new orthodoxy, which challenged the old, Gaelic and somewhat authoritarian one, or at least constructed an alternative lens through which it might be viewed, and presented that orthodoxy attractively interleaved with a wide variety of traditional, pop and on occasion even classical music. That new orthodoxy was not without its own questionable aspects, but television, and Byrne, created a dynamic where before there had been stasis, or at best tortoise-like movement. As one of Ireland's foremost cultural historians put it:

> The apparently unscripted nature of the show, which moved abruptly from levity to gravity, from frivolity to major social issues, without any warning, made it compulsive viewing for audiences. The paradoxical manner with which the programme combined a home-spun intimacy with "Brechtian" television techniques (displaying studio technology, the presenter himself calling the shots at times) gave it a place in life not unlike that of the provincial newspaper: people watched it if only because they were afraid they might miss out on something. (Gibbons, 1996: 79)

Another view was that the reason for the popularity of Byrne's programmes (from 1972 he also presented an influential morning radio programme) were related primarily to the social context, to a silence about issues that had become an all-enveloping fog. 'Surrounded by that silence, we wanted, in the 1960s, to hear ourselves speak in a charming, sophisticated and worldly wise voice' (O'Toole, 1990: 168).

The incremental, sometimes contested attempt to modify the context within which Irish social issues were discussed was not confined to news, current affairs and the *Late Late Show*, but extended into other entertainment areas, notably drama (Sheehan, 1987). One study tracing the development of Irish television drama from its inception in 1961 to the 1980s suggested that the long-running serial *The Riordans*, in particular, 'made deep inroads on a dominant ideology which looked on the family – and indeed the family farm – as the basic unit of Irish society' (Gibbons, 1984: 43). It has also been endorsed for the way in which it gave women a central and progressive role in television drama (O'Connor, 1984: 130). A related essay, focusing more sharply on the urban drama *Tolka Row*, although critical of the long-running serial's tendency to confirm rural stereotypes of city life, argued that it was 'an attempt at reinserting the missing discourse of the urban working class into Irish culture and, as such, can be seen as an important element in television's wider ideological project' (McLoone and MacMahon, 1984: 63). RTÉ, however, did not always represent a uniformly liberal agenda. In 1967, it incurred odium and ridicule in almost equal proportions after banning a ballad by The Dubliners, 'Seven Drunken Nights', because of its references to alcohol and adultery (NAI, Taoiseach: 98/6/83).

Many of these arguments, however, rapidly took second place to the overarching debate about television's role in the developing crisis in Northern Ireland, which emerged in 1968 and had, by 1969, erupted into political violence of an intensity and scale not seen for three decades. What happened was as unexpected for the media as it was for the political establishments in Dublin, Belfast and London. RTÉ, for example, had no permanent office or studio facilities in Belfast in 1968.

In May 1970, two ministers were dismissed from the cabinet in Dublin, charged with attempting to import arms for use by nationalists in Northern Ireland. Neither was convicted, but the government was deeply shaken, and the media excitement intensified. There was a stand-off between the government and broadcasters in June 1971, when the taoiseach, Jack Lynch, argued that it was not appropriate for the publicly funded RTÉ to broadcast interviews with members of an illegal organization, i.e., the IRA (DD, 24 June 1971). RTÉ responded that the programme by Liam Hourihane, which had been broadcast in the prime 1 p.m. slot on RTÉ on 17 June would be amended only if a written directive was received in accordance with Section 31 of the 1960 Broadcasting Act. No directive was issued. Not long afterwards, on 28 September, the current-affairs RTÉ programme *Seven Days* carried interviews with leading members of both branches of the IRA. Regarded by the government as a direct challenge to its authority, this prompted the minister for posts and telegraphs to issue the first directive under Section 31, on 1 October, instructing the authority to:

> refrain from broadcasting any matter that could be calculated to promote the aims or activities of any organisation which engages in, promotes, encourages or advocates the attaining of any political objective by violent means. (MacDermott, 1995: 31)

This directive, which by law had to be reissued annually, remained in force until January 1994. No member of parliament ever raised an objection there to its imposition or renewal. At this remove in time from the events concerned, such a lack of concern with a major restriction on freedom of speech is difficult to understand, but the historical context suggests a number of reasons. These include a fear among all politicians that the Northern conflict might spill into the Republic (as, on occasion, it did); strict party discipline; and an unwillingness to appear to be supporting the IRA. It is also noteworthy that the newspapers offered scant comfort to their colleagues in broadcasting. The *Irish Times*, for instance, suggested on 2 October, in the immediate aftermath of the government's action, that RTÉ's coverage of the Northern conflict had been marked by 'too much breathlessness', and even suggested that its lack of judgment might lead to staff changes.

The wording of the government directive, despite its apparent clarity, offered little in the way of concrete operational guidance to RTÉ. Its breadth – theoretically, it could have banned all reports of such organizations everywhere in the world, including for example the African National Congress in South Africa – was enhanced by a studied ambiguity, which may well have been designed with intimidatory intent. The RTÉ Authority sought clarification, which was refused. The government put on further pressure in June 1972, when it met the authority to complain about its broadcast of mute film of IRA members. Later the same year, the situation escalated when RTÉ broadcast on 19 November a radio report based on an interview by one of its reporters, Kevin O'Kelly, with a senior IRA figure, Sean Mac Stiofáin (who, as it happened, was not publicly named in the course of the report). The government demanded an explanation, and the authority responded that although they agreed that in retrospect the interview had been a lapse of editorial judgement, they felt it was an internal RTÉ matter. The government rejected this explanation as unsatisfactory, dismissed the authority in its entirety on 24 November, and appointed a new one. Subsequently, the IRA leader who had been interviewed was arrested and put on trial. RTÉ were forced to surrender the tape of the interview, but, in court, O'Kelly refused to identify the defendant as his interviewee and was sentenced to a prison term – later reduced to a fine – for contempt of court.

These events created widespread public controversy, including protests by journalists employed by RTÉ. One distinguished commentator, Basil Chubb, professor of politics at Trinity College Dublin, went so far as to call for news and current affairs to be separated from the rest of broadcasting 'under the control of an independent board that ... has the will and the courage to insist that Government interference be formal and public and to resist covert pressure' (*IT*, 31 Aug. 1972). Later, he was to observe that the increasingly lengthy list of instances of government involvement in the affairs of the national broadcaster 'were seen as a sinister catalogue of events by some, and could hardly be viewed with equanimity by anyone' (Chubb, 1974: 81).

The Northern situation had already destabilized the government, and in 1973, Lynch called a snap general election to secure his position. He succeeded in increasing his party's vote, but a decision by the two opposition parties to join forces for the first time since 1954 saw him lose office. The new minister for telegraphs was Dr Conor Cruise O'Brien, who had, since 1969, been a Labour member of the Dáil.

Mainstream national print media in the Republic experienced the late 1950s and early 1960s as a period of relative calm. The *Irish Times* was undergoing a process of retrenchment, especially with the closure of the *Evening Mail* and the *Sunday Review*, but board changes were paving the way for developments that would see that paper expand considerably towards the end of the decade. Douglas Gageby, a Dublin-born, Belfast-raised Protestant nationalist, who had played a key role in the establishment of both the *Sunday Press* and the *Evening Press*, accepted an invitation to join the *Irish Times* board in 1959. As a director, he attempted, without a great deal of success, to broaden the paper's appeal. Then in 1963 an unexpected editorial vacancy saw him assume full editorial control. For the next decade, he consciously moved to expand the paper's boundaries northward and westward, away from the Dublin metropolitan area, which at that time accounted for a large proportion of the paper's total sales. Major T.B. McDowell, the paper's managing director, facilitated this process by creating new financial systems within the paper, giving editorial executives much greater budgetary control.

Like RTÉ, none of the Dublin newspapers had a substantial presence in Northern Ireland in the early 1960s. The *Irish Times* broke new ground in this area, not only by reporting on the Northern Ireland parliament at Stormont at considerable length, but by sending writers to interview the younger generation of Northern Ireland's politicians, and to assess the situation.

This enlargement of the newspaper's agenda was complemented, geographically, by the development of a keen interest – and appropriate editorial coverage – in non-metropolitan Ireland, particularly the west. Intellectually and journalistically, it was accompanied by the development of a cadre of specialist writers. These were – unusually for the time – given their own bylines, and their work, focusing on areas like education, religion and women's affairs, to some extent reflected, and to some extent informed, the changes taking place in Irish society. It pioneered investigative features on controversial topics, most notably by Michael Viney, a gifted young English journalist. His extensively researched series on topics such as divorce, illegitimacy and adoption broke important new ground, and their promotion contributed substantially to growth in the paper's circulation.

Changes elsewhere were slower, but were occurring. At the *Irish Press*, both revenues and circulation were healthy, but there was an underlying sense of stasis, created by the presence of Eamon de Valera. As he prepared to leave political office and inaugurated his campaign for the presidency, which he was to fill for the next fourteen years, he was harried mercilessly in the Dáil by Noel Browne and a colleague, Jack McQuillan, the only members of the tiny National Progressive Democrats political party, who had examined the ownership structure of the *Irish Press*. In a withering attack timed to damage de Valera's campaign for the presidency, they accused him directly of a conflict of interest – a charge that stung, even in those relatively laissez-faire days (Coogan, 1993: 674–7). There was also, at the *Irish Press*, another malaise, mostly on the industrial-relations front, which boded ill for the future. For the time being, this was concealed by the fact that the Press group's management formed part of a consortium of Dublin newspaper managers, who met trade unions as a group and who operated a 'one out, all out' policy which meant that unions (mainly the printers' unions, the strongest in the industry) could not pick off employers one by one.

The strength of this arrangement was evident in the course of a major printers' strike in the summer of 1965, which closed the capital's newspapers. That strike was settled, partly because of the firmness of the employers, but also because of a factor that underlined the commercial role of newspapers in the city's life. The *Evening Press*, at that time, was the primary vehicle for full-page advertisements by major retail stores, in which the availability and prices of a wide range of items, from kitchen utensils to clothing, would be spelt out in great detail. This was primarily for the benefit of non-metropolitan readers, who used these advertisements as essential information on their carefully planned excursions to the capital.

At the height of the strike, a deputation of centre-city traders met the *Irish Press* management to explain that, since the strike had begun, their turnover had fallen by 30–40 per cent. The newspapers' managers explained that the cost of settling the strike on the printers' terms was prohibitive; the traders agreed to subvent the cost of settlement by increasing their display advertising by a guaranteed percentage for a period after the papers came back on the street.

The daily *Irish Press* also secured a new editor in 1968: Tim Pat Coogan. His appointment was unusual for a Fianna Fáil paper in that he was the son of a former Garda commissioner whose allegiance to the Fine Gael party had been well known. Coogan, who had served his apprenticeship on the *Evening Press*, was an editor of considerable dynamism. The founder of the *Irish Press*, Eamon de Valera, now president of Ireland, was no longer actively involved in running it. His son Vivion, who had taken over the reins, was also a backbench government member of the Dáil. The paper was still starved of resources and capital, running 16-page issues, compared to the 30–40 pages produced by the *Irish Times* and *Irish Independent* – but, editorially, Coogan had a measure of freedom, which he used to good effect. He enhanced the work of the paper's independent political correspondent, Michael Mills, by preventing sub-editors from shortening his reports unnecessarily; he employed a corps of effervescent young women journalists, some of whom – like Mary Kenny and Anne Harris – went on to greater heights in Irish and British journalism; and he encouraged his literary editor David Marcus' 1968 introduction of a regular page of new Irish writing, where many young writers who achieved fame in the 1970s and 1980s first cut their teeth.

The 1965 strike had seemed to represent a victory for the employers over the printers. But it had been bought at a high price and, when the printers again went on strike in 1973, the employers' line wavered. As it happened, this strike coincided with a general election. On the day the new government was formed, the *Irish Times* broke the 'one out, all out' agreement with the other newspaper managements, and appeared on the streets on its own. It published a front-page explanation for its decision, adding that it would not capitalize on the situation by printing additional newspapers. The next day, the *Irish Independent* and the *Irish Press* settled with the printers and followed suit.

This was to have long-term consequences, for the *Irish Press* newspapers in particular. Shorn of the protection of the other managements, and bedevilled by a long history of poor industrial relations, it was now in an exceptionally weak position to face the labour issues thrown up by the arrival of a new generation of newspaper technology. Although the drama and controversy of events in the North from 1968 were to sporadically enhance its circulation, this could not prevent the erosion of its readership and commercial bases by other factors.

To outside observers, the Independent group's papers appeared to change little during the 1960s, but the daily newspaper in particular was changing internally in ways that undermined its stereotype as a Fine Gael-supporting, conservative Catholic newspaper. Its editor, Frank Geary, was a newspaperman of the old school, in that comprehensiveness and accuracy were his two core values. Though somewhat indifferent to the new journalism of opinion and investigation developing elsewhere, particularly in the *Irish Times*, he gave some of his younger staff a relatively free hand, particularly in the areas of editorials and features: the feature department was effectively controlled by a small group of leader-writers, in a way that was structurally different from the other national papers, where editorial executives took specific responsibility for feature areas.

Its innovations in this area included a weekly survey of the continental European press, inaugurated in the wake of Ireland's first – unsuccessful – application to join the Common Market in 1963. It became probably the first Irish newspaper from the nationalist tradition to salute the memory of those Irishmen who had died in the First World War: one of its editorials, on the fiftieth anniversary of the outbreak of that war, consisted of the well-known poem 'For a Dream', written by the Irish poet and patriot Tom Kettle, shortly before he died at the Somme (*II*, 4 Aug. 1964).

In June 1965, it became the only national newspaper to editorially support the writer John McGahern, whose book *The Dark* had been banned by the censorship board. Although it was to some extent critical of the book, it commented:

What matters is that Mr McGahern has taken a most serious topic, the misery that assails an adolescent mind; he tackles it seriously, with sympathy and sincerity; he sets it in a framework which, we believe, undermines the value of what he has tried to do but which has no bearing whatever on the author's intention or his effect. (*II*, 3 June 1965)

In 1966, it mended its ideological fences not only by in April saluting the fiftieth anniversary of the Easter Rising (which it had roundly condemned at the time), but also by in May marking a significant break with its stance during the Spanish Civil War, by condemning Franco's repression of striking miners in the Asturias.

Geary was succeeded in 1968 by Louis MacRedmond, a young barrister who had been in charge of the editorial writers, and whose editorial stance embodied a more critical, though always loyal, attitude towards the Catholic church than those of his predecessors. In 1969, however, following a momentary downturn in the paper's profits, he was dismissed (technically, the board said they wanted him to assume other responsibilities, but there was little doubt but that a dismissal was what had actually occurred). Unusually, all his journalistic staff made a token protest by leaving the building during the production process and parading around the block, after he had dissuaded them from actually going on strike. Even more unusually for Irish journalism, his principal rival, the editor of the *Irish Times*, Douglas Gageby, immediately put him on a retainer, and published frequent articles by him until he secured permanent employment.

Outside the mainstream media, increasing economic activity, the new volatility in political life, and the wider availability of higher education was producing a potential market of literate young readers, and this was met at least in part with the publication of a fortnightly current-affairs magazine, *Hibernia*. The magazine had existed since 1937 as a monthly current-affairs journal aimed at the Catholic middle classes. In 1968, however, one of its contributors, John Mulcahy, acquired the paper from its proprietor, and turned it into a lively, irreverent and often well-informed magazine which specialized in an eclectic but highly marketable mix of political gossip and features, book reviews, and authoritative business and financial journalism. Its tone was crusading and investigative: by 1973 it was already carrying articles alleging conflicts of interest and possible corruption in relation to the activities of local politicians in the greater Dublin area – an issue which resurfaced, with dramatic effect, at the end of the 1990s. A combination of an expensive libel action in 1976, and an ill-judged expansion to weekly publication in the following year, finally forced its closure in 1981, when it had a circulation of approximately 17,500.

In Northern Ireland, the late 1950s and early 1960s saw the development of a greater interest, by both the BBC and ITV, in the internal political arrangements and circumstances of Northern Ireland. Towards the end of 1958, the BBC presenter Alan Whicker went to Northern Ireland to prepare a programme for the BBC current-affairs slot, *Tonight*. His televisual essay, when it was broadcast on 9 January 1959, presented to the national British audience a part of their own state with which they would, at best, have been only dimly familiar. It was not a flattering portrait, and although it caused little comment elsewhere in the UK, it evoked widespread unionist protest in Northern Ireland, particularly through the newspapers, which attacked the BBC for its slights, real or imagined, on Northern Ireland's character, people, and habits. Whicker had prepared further reports on Northern Ireland. They were never broadcast. Only three months later, however, controversy erupted again when the BBC broadcast a US series in which the American journalist Ed Murrow interviewed celebrities, including the Irish actress Siobhan McKenna. McKenna, who was born in the west of Ireland, had strong nationalist views (Cathcart, 1984: 194). Within days of 25 April the broadcast, Brian Faulkner, the unionist politician who was later to become prime minister of Northern Ireland, resigned from the province's BBC advisory council in protest. A scheduled showing of the second part of the interview with McKenna was cancelled.

This acute sensitivity to outside observers and broadcasters, incubated for decades in the less public, but equally sensitive environment of radio, was to be accentuated by television, and remained a feature of Northern Ireland politics and broadcasting for many years to come. Although the BBC did make attempts to broach contentious subjects it operated by and large in a manner designed to assuage rather than provoke unionist concerns. But the BBC no longer had the playing field to itself.

The first broadcast from the new Ulster Television Network, the ITV franchisee, took place on Halloween 1959. The new service could acquire viewers from only one source – the BBC – and it did precisely that, winning 60 per cent of the audience. This paralleled developments in Britain, where the more public-service oriented BBC found itself at a disadvantage against the profit-oriented and populist character of ITV. The competition between the two networks in Northern Ireland had a beneficial effect in that neither could ignore the need to build constituencies and audiences within each of the major religious and political traditions: UTV explicitly stated that it was in the business of community reconciliation (Cathcart, 1984: 187). The dynamics of the situation, however, were to ensure that, even with this added impetus, developments would be slow, and few major risks would be taken.

Relationships in broadcasting between Dublin and Belfast became more cordial, and a new telephone connection established between the BBC and RTÉ enabled them to share sports broadcasts as well as news and current affairs. The visit to Belfast in January 1965 by the taoiseach, Sean Lemass, to meet the Northern prime minister, Terence O'Neill, gave rise to a flurry of cross-border meetings between civil servants, politicians and – inevitably – BBC and RTÉ personnel.

In 1966, the fiftieth anniversary of the 1916 Rising in Dublin was marked, in Northern Ireland, by an outbreak of nationalist sentiment and by the display of nationalist flags (illegal under legislation passed by Stormont). BBC and UTV bulletins reporting these events were seen by unionists as evidence of network bias. Not even a visit to Belfast by Queen Elizabeth, evidently designed to reassure unionists, could calm the situation.

The emergence of the civil-rights movement in Northern Ireland in 1968 created a new situation, in which the media in general, but particularly television, played a new and central role. There had been sporadic disturbances in Northern Ireland for many years, but the political convention that Northern Ireland affairs could not be discussed in the Westminster parliament meant that nationalist grievances lacked an effective sounding board (the Stormont parliament was effectively under the permanent control of their unionist opponents). Television now became that sounding board, amplifying both the protests against the political status quo in Northern Ireland, and the unionist government's ham-fisted and often brutal attempts to suppress them. This was particularly true of dramatic footage secured by an RTÉ cameraman, Gay O'Brien, of a civil-rights march in Derry in October 1968 that was met by water cannon and baton charges. When the BBC used this footage in preference to their own, television itself became an extension of the battleground, as both sides jousted for control of the medium.

Particular exception was taken to BBC reports of serious civil disorder in Belfast in August 1969, when whole streets of nationalist homes were torched by loyalist vigilantes. The BBC reported what had happened without identifying the source of the attacks, and subsequently defended its decontextualized report. Eight years later, Martin Bell, who had been seconded from London to join the BBC team in Northern Ireland, charged that this had been a grave mistake and was 'the only time when I was stopped by the powers above from saying what I wanted to say' (Cathcart, 1984: 212).

Northern Ireland broadcasters shared one dilemma with their counterparts in the Republic – at least until the issuance of the directive under Section 31 of the Irish government's Broadcasting Act: how should paramilitary forces be represented? The BBC showed a number of films of IRA personnel on both sides of the border (loyalist paramilitaries were, until 1974, a much less potent reality). This led to considerable controversy, and even – in April 1971 – to the imprisonment for four days of the BBC reporter Bernard Falk for refusing to identify a defendant charged with IRA membership as someone he had previously interviewed for television. This in turn produced a system of referring such critical questions upwards to higher authority.

These events left their mark in the print media in Northern Ireland too, but here developments in editorial policy were being paralleled, at least in the cases of the *Irish News* and the *Belfast Telegraph*, by changes in ownership and direction. In 1964, Dr Daniel McSparran, who had controlled the *Irish News* for three decades, retired. His place was taken by his son, also named Daniel McSparran. Two years later, the paper had a new editor, Terence O'Keeffe, a Liverpool-born, Jesuit-educated journalist who nudged the paper away from its Catholic bulletin-board image towards an openness to other traditions (Phoenix, 1995: 35). Its editorials, nonetheless, mirrored both the rising hopes and the subsequent disillusion of Northern nationalists when the promise of 1965 gave way to the conflicts of the late 1960s. On 10 October 1968, in the aftermath of the civil-rights march in Derry, one editorial concluded that unionism would never really change its spots, and that the nationalists, in accepting the role of official opposition at Stormont (as they had done in 1965 at the urging of the Dublin government), had made a fatal political mistake.

The spread of civil conflict helped the circulation of all Northern Ireland newspapers, as each side consumed news and information avidly. There was a growth in political cartooning, in which a divide rapidly appeared between UK-based and some Irish media (Darby, 1983). Whereas Irish-based cartoonists could occasionally rise to a sophisticated, sometimes black humour, UK cartoonists in the mainstream (and especially the tabloid) press:

> tended to avoid those aspects of the Ulster issue which would normally emphasise the differences between the political left and right ... it was

tempting to regard events in the province as a struggle between the British virtues of reason, decency and moderation and the Irish antitheses of these. (Darby, 1983: 120–1)

All the papers had now been modernised to a considerable extent: when the *News Letter* put news on its front page for the first time in 1961, it was the last of the three Belfast daily papers to do so. And the deepening divisions did not preclude some commonality of interest: in 1969, when an unfounded rumour circulated suggesting that the BBC broadcasting operations in Northern Ireland were about to be closed down because of a shortage of money, not only the *Irish News*, but the *Belfast Telegraph* and the *News Letter* published editorials condemning the idea (Cathcart, 1984: 211).

As far as offering an analysis of the Northern Ireland problem was concerned, the *News Letter* did its best to suggest that the cause was less unionist intransigence or discrimination and more a desire, on the part of militant republicans, to recapture the political support that had been seeping away from them since partition in 1921. The *New Letter* pointed out that although in 1921 more than 30 per cent of the electorate in Northern Ireland had voted for nationalist candidates, between 1929 and 1969 that percentage had been halved (Shearman, 1987: 47). These arguments did not impress more militant elements within nationalism. During the 1970s, there were nine bomb incidents at or in the vicinity of the newspaper's offices. One bomb in 1972 injured twenty-one of the company's employees, and another in 1973 damaged a strongroom containing bound volumes of the paper going back to 1738.

Risks of a different kind affected the *Belfast Telegraph*. As already noted, the paper's management had been wrestling with the problem caused by the untimely death of its managing director, Bobbie Baird, in 1953. Death duties on his estate (which included the *Telegraph*) were enormous, and although the directors lent money to cover this debt, it became evident that the paper itself would have to be sold. It was eventually purchased by the Canadian newspaper magnate Roy Thomson in September 1961 for £1.25 million. The closure of the *Northern Whig* in September 1963, in the wake of a strike by the National Union of Journalists, cleared the field of at least one competitor (although the *Whig* was a morning paper). Thomson's canniness in making the purchase was richly rewarded in subsequent years: profits on the *Belfast Telegraph* approximately covered his losses on the London *Times*, which he had purchased earlier.

With the *Telegraph*, of course, he had also inherited Jack Sayers, one of the most influential – and disappointed – editors of his era. Over most of the following decade, until his retirement in 1969, Sayers supported moderate unionism in its unsuccessful attempt to find common ground with moderate nationalism, and did so with a single-mindedness that sometimes blinded him to the power of the forces ranged against it. As Ian Paisley became part of the movement aimed at unseating Terence O'Neill in 1966, Sayers wrote starkly about the dilemma facing Northern Ireland in his editorial of 12 July: the choice was 'between responsible government through the present unionist leadership or a form of dictatorship through a religious war led by a latter-day "Mad Mullah"' (Gailey, 1995: 108). Throughout 1967 and 1968 he worked in public and private to bring about a rapprochement between the leaders of both communities.

The media-consumption habits of nationalists and unionists in Northern Ireland remained sharply divided. In 1970, 87 per cent of the *News Letter's* readership was Protestant and 93 per cent of the *Irish News'* readers were Catholic. The *Belfast Telegraph's* readership was 68 per cent Protestant and 32 per cent Catholic, reflecting the breakdown of the two main allegiances in the province. Interestingly, the *Belfast Telegraph*, despite its editorial stance, did not alienate the more extreme supporters of either side. Up to half of its Protestant and Catholic readers supported a loyalist or strongly republican viewpoint respectively. Only 5 per cent read a newspaper edited in Dublin, and 32 per cent read a British paper. In spite of frequent criticism, the public broadcast media were thought to be the most reliable source of news by twice as many people as awarded that role to print media (Rose, 1971: 343–5). Catholics were more likely than Protestants to read a UK-published paper, probably because of their coverage of sport and entertainment than for political reasons. It is also relevant that although UK tabloids were noted for their anti-republican and pro-British army attitudes for the duration of the conflict, a certain level of localization ensured that the more jingoistic headlines were confined to the editions circulated on the larger island.

The more extreme unionists, against whom Sayers had been inveighing, now even had a paper of their own. This was Revd Ian Paisley's *Protestant Telegraph* (1966–82) – the name itself undoubtedly a backhanded compliment to the *Belfast Telegraph* – which was founded in April 1966 and which provided its readers with a potent mixture of . anti-Catholic invective and proto-unionist political sentiment. Paisley's Free Presbyterian church, of which it was effectively the mouthpiece, was to grow substantially, both in the number of its adherents and in the resources it had available. The *Protestant Telegraph* had a circulation of about 15,000 at its peak.

But the era also produced a novelty of a different kind in the shape of *Fortnight*, a magazine that was established in 1970 by a group of young Queen's University lecturers and others working to accommodate community differences through rational discussion. The name *Fortnight* was chosen because it was thought to be totally free of any sectarian or political undertones. Its first editor, Tom Hadden, was a lawyer, and this may have been responsible for an emphasis, in the early years of its existence, on legal and civil-rights issues. Its even-handedness was not always welcome. When it opposed internment without trial in 1971, its printers in Northern Ireland refused to handle it, and it was printed in Dublin for three months until a grant from the Rowntree Trust enabled it to buy its own press. After going through various vicissitudes, not the least of which were financial, it established a rare reputation as a journal in whose pages politicians and writers of every political persuasion could find a home. Despite its name, however, by the twenty-first century it appeared only monthly. It finally ceased publication in 2012, wounded both by declining circulation and advertising revenues but also – ironically – the calmer political conditions then prevailing.

6 Crossing a Watershed, 1973–85

The decade or so from 1973 onwards saw major changes in Irish media, particularly in the Republic. The North was more immune to major structural and ownership changes for some time after the purchase of the *Belfast Telegraph* by the Thomson organization, but even here the decade saw the emergence of new titles and the sharpening of competition. In Dublin, the somewhat old-fashioned and complacent ownership and management systems were being reformed or broken up as a more aggressive and modern form of capitalism took shape. New titles were appearing; the influence of UK media was becoming a subject of debate, particularly in relation to their treatment of the Northern Ireland issue; and the evolution of broadcasting reflected not only the all-pervasive influence of the Northern conflict but recurrent controversies about national identity, programme choice and political interference.

By early 1973, the economy of the Republic had been on an upswing for some years. Even the *Irish Times'* balance sheet had improved under the influence of a number of variables: economic growth, political controversy and venturesome editing. A number of significant developments then took place in rapid succession. On 2 March, Major T.B. McDowell, who had been chief executive of the company for eleven years, also became chairman. Twelve days later, the newspaper – then in the middle of a printers' strike – broke the 'one out, all out' agreement between the Dublin newspaper managements to publish news of the formation of the new government the previous day. It did so, it said, because 'the Directors consider that they have an obligation to the public which overrides all other considerations' (*IT*, 14 Mar. 1973). Simultaneously, three of the five directors who controlled the voting shares in the company were coming to the conclusion that they wanted to retire. They had received little or nothing by way of dividends for years, and indeed had had to borrow to invest further in the paper from time to time. The shares were now acquiring some value – so much so that it would have been, in practical terms, difficult for the other two directors to purchase them all prior to reconstituting the board. There were also rumours that Roy Thomson, already the owner of the *Belfast Telegraph*, might be interested in adding to his Irish portfolio.

This possibility was averted by a major restructuring of the organization devised by McDowell. A trust was set up on 5 April 1974, which acquired the

shares of all the directors on the basis of borrowings of more than £2 million
from the Bank of Ireland. This trust was given the power to appoint a
majority of members of the board of the company that now controlled the
Irish Times directly. Douglas Gageby, who had been a director as well as editor,
retired as editor.

Changes of a similar magnitude, although not of the same kind, were
taking place simultaneously at the Independent newspapers. The company
was still controlled by the Murphy and Chance families, and for some time
rumours had been circulating to the effect that the Independent group was
ripe for a takeover. The country's first financial weekly, *Business and Finance*,
declared on 1 March 1973 that it was 'grossly undervalued'. Its pre-tax profits
in 1972 were over £750,000, and it had been extending steadily and profitably
into the provincial-newspaper field. On the east coast, where the population
had, generally speaking, a higher disposable income and where more
commercial activity took place, it had acquired a number of titles in a wide
swath running from Drogheda in the north to Wicklow in the south, as well
as the *Kerryman* in the south-west, one of the largest and most prosperous of
all Irish provincial papers.

The *Kerryman* had fallen into the *Independent*'s hands on the death of its
long-time owner, Dan Nolan; the Independent group was to suffer a similar
fate as its founding shareholding families began to lose interest in continuing
the business. Speculation about who would purchase it at first centred on the
wealthy McGrath family, which had built an industrial and business empire
on the foundation of a franchise, which they had been granted by an accom-
modating government in 1930, to run a national sweepstakes to benefit Irish
hospitals. On 21 January 1973, however, this possibility evaporated in the wake
of the publication, in the *Sunday Independent* of that date, of one of the most
remarkable pieces of investigative journalism that had ever appeared in an
Irish newspaper up to that time and indeed for many years afterwards. This
was an exposé by the journalist Joe MacAnthony of the activities by
McGrath's sweepstakes company, many of them illegal, in the foreign jurisdictions in which sweepstakes tickets were sold.

The article, which ran to some 8,000 words and over three pages of the
paper, had been planned originally as a two-part series. The editor of the
paper, Conor O'Brien, realised, however, that if the first half was published,
the second half would in all probability never see the light of day; accordingly, he decided to run it all in one article. The McGraths were incensed, and
withdrew all their lucrative advertising from the group for some weeks.

MacAnthony resigned not long afterwards, after it had been made clear to him by a regretful O'Brien that his career within that organization had effectively reached a full stop, at least in terms of promotion. Other potential purchasers began to eye the company.

On 17 March 1973 it was announced that a young Irish financier and businessman, Tony O'Reilly, had taken a large stake in the company, together with an associate of his named Nicholas Leonard, a former financial editor of the *Irish Times*. O'Reilly's initial purchase amounted to no more than 20 per cent of the controlling shares, but he persuaded the Independent board, of which he was now a member, to agree to a performance-related formula through which he would be permitted to increase his equity holding in the company if it achieved certain specified commercial targets. In this way, he increased his equity incrementally. By 1980, when he was appointed chairman of Independent Newspapers, his initial investment of about £1 million was worth more than £5 million.

These changes, none of which had been predicted, threw the normally placid world of Irish journalism into turmoil. The changes in share ownership at both newspapers had taken place against the background of the possible introduction, by the new government which took office in 1973, of a capital gains tax, which would have sharply reduced the windfall profits from any sale of company shares by the founding family. Such a tax, which had been mooted publicly during the general election campaign, was introduced by the government later the same year, but after the titles had changed hands. Alarmed by the suddenness of the change at the *Independent*, the group's 200 journalists took emergency action, effectively occupying the premises until they received a management guarantee that their jobs were not in jeopardy.

Even these difficulties were to be rapidly overshadowed by other, more global factors. In the middle of 1973, the Organisation of Petroleum Exporting Companies (OPEC) began relentlessly to increase the price of crude oil. Ireland, with its small, open economy, was more than usually vulnerable to the resulting downturn in the world economy. As the price of oil and oil-related products soared, manufacturing went into recession, consumer confidence evaporated, and Irish inflation escalated – to a rate of 26 per cent in 1976, the worst year of the crisis.

All these factors led to a collapse of the media-advertising market, particularly in the newspapers (with restricted budgets, many advertisers preferred to concentrate their resources on television and radio). This collapse in advertising revenue reduced both the size and quality of newspapers, as editorial and production budgets were in turn affected. The *Irish Times'* reserves dipped to dangerous levels.

The Independent group was experiencing similar problems, but was better equipped to deal with them. For one thing, it had a strong national base. For another, it began to look outside Ireland for profitable investment opportunities. O'Reilly's international connections – he became chairman of Heinz in 1973, and joined the board of Mobil Oil in 1979 – prompted a policy of diversification in media-related industries in Britain, the United States, Germany and Mexico. By the end of the 1970s, more than half the group's profits were being generated by subsidiaries or from investments located outside Ireland.

In addition, a number of editorial changes were beginning to reap dividends in terms of circulation. Vincent Doyle, the editor of the daily *Irish Independent* from February 1981, brought to his position a tough – some would say ruthless – approach to news, and a determination to beat the opposition on all major stories. This was allied to changes in layout, typography and design that made the *Irish Independent*, although a broadsheet, assume at least some of the display characteristics of a tabloid. Its front page became more and more striking, and this hybridization of style not only helped it to fend off, to some degree, the competition from imported UK papers, but also possibly attracted younger readers away from the more typographically staid *Irish Press.* The *Sunday Independent*, edited from January 1984 by Aengus Fanning, had adopted different tactics, but with similar results. Its new emphasis on pungent opinion columns, gossip and fashion enabled it to cut the circulation lead of the *Sunday Press* from 100,000 to 30,000. It was to overhaul it finally in 1989.

If the Independent group had a firm grip on middle Ireland, buttressed by an increasing range of profitable foreign investments, the economic and readership base of the Irish Press newspapers was less secure. Although it appeared to survive the mid-1970s relatively unscathed, this was in large part because the Fianna Fáil party was in opposition between 1973–7. However, readership of the Irish Press group's papers (with the possible exception of the *Evening Press*), though substantial, was beginning to assume characteristics that did not generate much commercial enthusiasm. The readers – by contrast with those of other papers – tended to be older, were more likely to be male, to live in non-urban environments, and to have relatively smaller amounts of disposable income.

Nonetheless, in 1980, its last good year, the company reported profits of

£715,000 before tax. The *Evening Press* was outselling the *Evening Herald* by 30,000 copies a day, with a total of 172,000; and the *Sunday Press* had 55 per cent of the domestic Sunday newspaper market with a circulation of just under 400,000 copies a week. The daily *Irish Press* was selling a respectable 100,000.

The death of Vivion de Valera in 1982 meant that control passed to the founder's grandson, Dr Eamon de Valera, and this change was accompanied by management plans to make the long-awaited technological changes which were necessary if the group was to meet the challenges of the powerful Independent group with any confidence. The attempts to introduce these changes were accompanied by mounting financial losses and sharply deteriorating staff relations. In the summer of 1983 the titles were off the streets for three weeks. This was only the first in a series of similar disputes. Mounting losses – £1.5 million in 1983 and £3.5 million the following year – prompted the drastic management decision to switch from hot-metal to computerized typesetting in 1985 without first securing union agreement.

One of the few initiatives to run counter to the general trend of retrenchment during this period was the establishment a new Irish tabloid Sunday newspaper, the *Sunday World.* The market for broadsheet Sunday newspapers in the Republic has not been a fiercely contested one. The British tabloid newspapers, however, were virtually unopposed in the Sunday sector in Ireland: the *Sunday Independent* and the *Sunday Press* were broadsheets that attempted, not always successfully, to stretch their appeal to cover both the popular and quality ends of the market.

The *Sunday World* was launched on 25 March 1973 to fill the tabloid gap in the range of Irish Sunday titles. Its founders were two businessmen, Hugh McLoughlin and Gerry McGuinness, partners in the Creation Group, which published a number of women's magazines and had a relatively sophisticated colour press. McLoughlin, the son of a Co. Donegal stationmaster who had built a business career in Dublin, had an extraordinarily influential role in the creation of new Irish media. As well as his involvement in the Creation Group, and the *Sunday World,* he was behind the publication of the first Dublin free sheet, the *Dublin Post,* and had been closely involved with the restructuring of the *Irish Farmers Journal* in 1950 (it had been established originally in 1948, but had limited success in its first two years and sometimes suffered breaks in publication). He had also founded *Business and Finance* in 1964 with Nicholas Leonard, up to then the financial editor of the *Irish Times.*

The *World* was a publication that had absorbed many insights from its UK competitors, notably a cheeky willingness to engage in sexual innuendo: its marketing slogan was 'We Go All The Way'. It was also in many ways distinctively Irish. It promised, and indeed occasionally delivered, tough and fearless journalism. Its first editorial noted: 'In this newspaper, common objects like spades will be known as spades ... They are used to dig with and we intend to dig' (*SW,* 1 Mar. 1973). It realized that coverage of television was popular (the existing newspapers tended to ignore their electronic rival). It adopted a campaigning mode, including collecting money for deserving causes. It had a problem page, the first of its kind in Ireland. And it specialized in short paragraphs, screaming headlines and huge bylines for journalists.

In spite of the fact that it sold initially for six pence – a penny dearer than both its Irish competitors – it was a success virtually from the start. It had initial capital of only £40,000, on which it was thought it might survive for six weeks, but by December 1973 its sales were at more than 200,000, and by 1975 its annual profit was £100,000.

Insofar as it had a political line, it was soft-focus Fianna Fáil. This party had the greatest following among the urban working classes and among small farmers. The *World*'s appeal, however, was not confined to these groups, and it rapidly acquired a readership that was not only geographically widespread – it was as popular in rural as in urban areas – but whose sociological profile spanned all social classes, making it uniquely attractive to advertisers. It also made substantial inroads into Northern Ireland, which had no locally published Sunday newspaper at all. Its Southern origins did not, however, go unnoticed. In May 1984, its Northern editor, Jim Campbell, was the victim of an unsuccessful assassination attempt by loyalists. Though seriously wounded, he returned to his desk some months later. In 1988, at the height of the *World*'s circulation, some 96,000 of its total sales of 355,000 copies were in Northern Ireland.

By then, however, it had changed ownership. In 1977, the fortunes of the Creation Group were wavering, as the magazine market suffered, like others, from the long-running economic crisis. McGuinness and McLoughlin toyed with the idea of enhancing their revenues by publishing a tabloid evening paper – the *Evening Herald* and *Evening Press* were both broadsheet. In January 1978 O'Reilly and the Independent board, partly in response to this threat but also because they sensed an opportunity, made a successful offer of £1.1 million for McLoughlin's 54 per cent shareholding. This was followed, in April 1983, by the purchase of McGuinness' shareholding for almost £3 million.

By 1984, the contribution of the *Sunday World* to Independent group profits was already approaching £1 million, even after allowing for interest on the borrowings taken out to acquire the paper. Occasionally denounced from pulpits, it softened its campaigning stance as it was hit by increasingly expensive libel actions (in the latter part of the 1980s, its legal costs were averaging over £500,000 a year), but maintained a robustly laddish approach to its journalistic agenda. When its editor, Colin McLelland, was summoned before the Oireachtas committee on women's rights in 1985 to answer allegations of sexism, he refused, and sent the committee a letter noting:

> My own view is that there is a considerable difference between men and women. They look different, for a start, and I firmly believe that a tractor adorned by a girl in a bikini is infinitely more pleasing to the human eye than the same tractor adorned by a man in Y-fronts. To me, that's the way that God planned it. (*IT*, 18 July 1985)

Others attempted to follow in McLoughlin's tracks, but with less success. Tony Fitzpatrick, who was a former Dublin correspondent for the British Sunday newspaper *The People*, and who had subsequently occupied a number of senior editorial positions on the *Sunday World*, persuaded a number of investors to finance a new Sunday newspaper, the *Sunday Journal*, which would be aimed at the farming community, and would compete with the *Farmers Journal*, which published on Thursdays. This development was significant in that it involved the first major investment from non-Irish sources in an Irish national newspaper: a number of the investors, including the British financier Robert Holmes à Court, were based in the Isle of Man.

Taking on the *Irish Farmers Journal* seemed like – and, as events proved, was – a high-risk enterprise. Investors, alarmed at the rate at which the initial capital was being consumed, tried unsuccessfully to remove Fitzpatrick. He then turned the tables on them by persuading an Irish entrepreneur, Joe Moore, to buy into the paper as the original investors departed. Moore, a craggy and despotic businessman, assumed control with great enthusiasm, but the *Sunday Journal* went into rapid decline, with an increasing proportion of its decreasing number of pages being devoted to motoring matters, until it expired on 20 June 1982 with a circulation of some 20,000.

Emboldened by his success with the *Sunday World*, the entrepreneur Hugh McLoughlin decided on yet another foray into newspaper publishing and, in 1980, went into partnership with the former editor-proprietor of *Hibernia*,

John Mulcahy, and a major printing group, Smurfit, to set up a printing operation and to produce a quality Irish Sunday newspaper, aiming at a perceived gap in Irish journalism. The *Sunday World* was a tabloid; the *Sunday Independent* and *Sunday Press* were established middle-market papers; the circulation in Ireland of British Sunday newspapers such as *The Observer* and *Sunday Times*, though small, indicated that there was another market here which might also be tapped.

The *Sunday Tribune*, accordingly, was launched in October 1980, with an energetic and talented ex-*Irish Times* journalist, Conor Brady, as its editor. Initially at least it seemed as if this had been yet another canny gamble. Its 1981 audited circulation was 110,000, certainly enough to make the enterprise viable, and its readership was over 400,000. In May 1981, however, there was a restructuring of the companies involved, in which McLoughlin ceded his part-ownership of the printing operation to Smurfit, and Smurfit in turn ceded to him their share of the ownership of the *Sunday Tribune*. Printing and publishing were now two entirely separate operations, a development that was to have considerable significance for the future.

Initially, the *Sunday Tribune* went from strength to strength. It developed a particular expertise in political reporting at a time when Fianna Fáil, the country's largest political party, was undergoing a leadership crisis. The *Tribune's* political correspondent, Geraldine Kennedy, was a young journalist who had come into national journalism from her local paper in Waterford, and who quickly established a major reputation for herself as the author of a number of political exclusives. Her activities, and those of another talented and independent-minded journalist, the *Irish Independent's* Bruce Arnold, were focused on, among other things, Fianna Fáil's leadership crisis. From 1979, the party's new leader was C.J. Haughey, one of the two ministers who had been dismissed from the cabinet in 1970 and subsequently arraigned on a charge of importing arms illegally – a charge of which he had been acquitted.

It emerged in 1983 that while Haughey had been head of government, the telephones of both Kennedy and Arnold had been illegally tapped in an attempt to find out where they were getting information damaging to Haughey's interests. Both sued the state for compensation, and won. And the Arnold and the Kennedy cases established the important precedent that the constitutional right to privacy extended to the telephone (Arnold, 1984: 2). Kennedy would in 1987 go into politics directly, as a deputy for the Progressive Democrats, but would lose her seat at the following election and return to journalism.

The modest but genuine success of the *Sunday Tribune* could now be contrasted with the fate of yet another gamble by McLoughlin. He had identified what he saw as yet another gap in the market – a gap that could be filled by an Irish tabloid paper. Accordingly, after a three-month planning period, the *Daily News* was launched on 7 October 1982.

The new paper was effectively a subsidiary of the *Sunday Tribune*, which was to be paid some £40,000 a week for printing it – money the *Tribune* would owe to the printing company now controlled by Smurfit. The effect of this operation was to expose the fatal under-capitalization of the whole enterprise. Initial enthusiasm at the *Daily News* was rapidly transmuted into panic, the layout became more and more improvised, and the stories and pagination more and more flimsy. It eventually collapsed after only 15 issues, and its 50 editorial staff were given notice. The collapse of the *Daily News* helped, in turn, to drag down the *Sunday Tribune*, which also went into liquidation in October 1982, crippled by *Daily News* debts. Not all the *Tribune*'s losses were attributable to the *Daily News* experiment: the *Tribune* had also launched a very expensive colour supplement, which proved a major drain on its resources.

Unlike the *Daily News*, the *Sunday Tribune* still represented an attractive proposition for potential publishers. Its title was acquired from the liquidator for a nominal sum in November 1982 by a company called St Stephen's Green Publications, two of the principal shareholders of which were a journalist, Vincent Browne, editor of a the current-affairs magazine *Magill*, and Tony Ryan, a businessman who was at this stage in the process of making a fortune (much of which he was later to lose) in an airline-leasing company called Guinness Peat Aviation. The paper was relaunched on 17 April 1983 with Browne as editor, and with a staff that had been cut to 32.

Although under Browne's editorship the paper rapidly acquired a reputation for news exclusives and outspoken comment on public issues, his somewhat erratic management style, and poor staff relations, began to create problems. In 1984 Ryan put his shareholding up for sale, and it was acquired by other businessmen who were not as in tune with Browne and his objectives as Ryan initially had been. By April 1986, the shareholders had decided to refinance the company and, in October of the same year, the *Sunday Tribune* was reorganized into two companies, one of which, it was hoped, would bolster the organization's revenues by publishing a series of supplements designed to cash in on the growing market, particularly in Dublin, for free local newspapers. It was a brave new dawn, but it was to plunge the paper into even deeper financial difficulties.

The seesaw fortunes of these relatively new titles bore witness to the fact that, despite the difficult economic situation, entrepreneurial optimism was not in short supply. Commercial success, however, was more marked in the magazine field. Here, as well as *Business and Finance, Magill,* which had been founded in 1977, outlasted all its competitors, and for a considerable time managed to finance its unique blend of investigative reporting and powerfully original writing by virtually cornering the market in high-gloss colour advertising. Browne, its founder and sometime editor, had cut his journalistic teeth on the *Irish Press,* and displayed a rare verve that – although it sometimes ended by alienating talented young journalists – also ensured that *Magill* remained a powerful and influential voice.

Apart from the fortunes of the individual titles, newspapers in the Republic experienced, during this period, a number of structural and professional problems. Structurally, newspaper managements had begun to pay increasing attention to government–press relationships, not in the area of censorship, but in the area of government support for industry. One international study (Picard, 1983: 16), which looked at government intervention in the economics of newspaper publishing across sixteen advanced Western democratic countries, found that Ireland, with a weighted score of 7.4 (as against Sweden's 19.2), came second lowest in the list of state subsidies, ahead only of Switzerland.

These and related matters had been of increasing concern to the Dublin Newspaper Managers' Committee, an organization that had existed for about half a century, and in which the main national newspapers co-operated on industrial-relations matters. By the early 1980s, the loss of advertising to television, and a perceptible reduction in newspaper circulations generally, led to joint action by members of the committee to promote newspaper advertising. This was a novelty: the national newspaper managements were traditionally suspicious of, or even hostile to, each other in areas other than the necessary co-ordination of industrial-relations policies. The new spirit of co-operation, however, led to the formation in 1985 of a new organization, National Newspapers of Ireland, which replaced the DNMC, but had a far wider range of functions, not least representing the industry and lobbying for its objectives. It has traditionally been concerned about four major issues: VAT on newspapers, particularly in the light of intensive competition from foreign titles; the defamation laws; taxation other than VAT; and press regulation generally. Its first major success was in 1984–5, when it succeeded in persuading the government to reduce the rate of VAT on newspapers from

23 per cent to 12.5 per cent. By 2016, it had been reduced to 9 percent, but was still one of the highest in Europe.

Campaigns to reform the defamation laws have met with less success to date. In June 2000, the *Irish Times* announced that it was increasing its financial provision for libel costs and damages in that year to £1.4 million from the previous year's £1 million (*IT*, 2 June 2000) – but the owners' campaign won two other major commercial victories. The first was in the 1992–3 financial year, when it secured a high court victory against the revenue commissioners ensuring that newspapers, as part of the manufacturing industry, were taxed at 10 per cent (corporation profits tax generally was 40 per cent). The second was in 1994, when NNI successfully prevented Cablelink, the principal supplier of cable TV services in Dublin, from subsidizing the cost of providing its subscribers with satellite services by slotting in local advertising.

The establishment of NNI was an initiative undertaken against a background of apparently waning interest in newspapers generally as a source of information and commentary on current affairs. One survey (MRBI, 1983: 14–15) noted that only 17 per cent of those who expressed an interest in current affairs chose newspapers as their prime source of information, as against 53 per cent who chose television, and 20 per cent who chose radio. The 17 per cent who chose newspapers were very noticeably middle class, and were to be found principally in the 25–64 age group. The same study identified three main areas in which media generally were said to have a major source of influence on people's thinking: aid to the third world, the nuclear threat and Northern Ireland. On other issues (marriage and family life, abortion, divorce and general issues) media ranked substantially below the family and church, and ahead only of politicians.

The relationship between the politicians themselves and the media was also evolving during this period. Three developments, in particular, marked this process. The first occurred in 1973, when the first non-Fianna Fáil government in sixteen years came to power. That government substantially reorganized the over-bureaucratized structure of the government information bureau (it later became the government information service), appointing as its new head a talented young television producer, Muiris Mac Conghail, who had been in charge of the controversial 1969 RTÉ television exposé of moneylending in Dublin. Under Mac Conghail, the media management of government became much more professional and efficient

The second major development in this area occurred in 1976, when the conflict in Northern Ireland was showing every sign of beginning to spill over into the Republic. Deaths from paramilitary or security-force activity in the North were occurring at the rate of one a day; there were large parades in Dublin in support of the Provisional IRA; and, in July, the British ambassador to the Republic, Christopher Ewart-Biggs, was killed by an IRA bomb near his home in Co. Dublin. Cabinet ministers were under severe pressure, both politically and personally. Members of their families received frequent telephone threats, and their personal security was increased to extraordinary levels.

This process culminated, in the summer of 1976, in the drafting of a proposed amendment to Section 31 of the Broadcasting Act to curb, insofar as possible, IRA activity and support in the Republic. This included a draft Section (3) that read:

> Any person who expressly or by implication, directly or through another person or persons, or by advertisement, propaganda or any other means, incites or invites another person (or persons generally) to join an unlawful organisation or take part in, support, or assist its activities, shall be guilty of an offence and shall be liable on conviction on indictment to imprisonment for a term not exceeding ten years.

The thinking behind this draft legislation became dramatically clear later in September when Dr Conor Cruise O'Brien, the minister for posts and telegraphs since 1973, entered the fray. A former diplomat and university professor who had become a member of parliament in 1969, O'Brien was an opponent of censorship, and had vigorously criticized the previous government's decision to dismiss the previous RTÉ Authority; but he was also sceptical about journalists' attitudes towards republicanism. Now he volunteered the view, in an interview with the *Washington Post*'s Bernard Nossiter, that he would not be averse to prosecuting the editor of the *Irish Press*, Tim Pat Coogan, for frequently publishing readers' letters that in his view offered aid and comfort to the IRA campaign. Not long afterwards, he added fuel to the flames by confessing that, for him personally, 'any coverage of Sinn Féin activity is too much' (Akenson, 1994: 422–3).

Politicians of all parties – including his own – reacted angrily. So did trade unions, not least the NUJ. The offending section was modified so that newspaper coverage would be excluded from its provisions: it was the price of saving the rest of the legislation.

The third development was a growing interest, on the part of the media, in the private lives of political leaders. In the 1970s and early 1980s, the politician–media relationship was marked – as it still is to a substantial extent – by a degree of media reticence uncommon in the United Kingdom. In October 1979, however, the president of Ireland, Dr Patrick J. Hillery, took the extraordinary step of inviting senior Irish journalists to his office to counter rumours that he was about to resign his presidency for personal reasons. The subtext was the suggestion that his marriage was breaking up and that he was intimately involved with another woman. As one historian – at the time a journalist dealing with the story – later observed:

> Confusion reigned in Dublin newsrooms as editors wrestled with what was on and what was off the record. A strong tradition in Irish journalistic life was being undermined: the private life of a President was not the subject of journalistic comment. (Keogh, 1994: 342)

Although it was ostensibly off the record, accounts of this extraordinary event, varying only in the amount of detail each contained and sourced unmistakeably to the president himself, all appeared in the media on the following day, and the rumours were never heard of again.

Later politicians were not to be so fortunate. Emmet Stagg, a junior government minister in the 1992–4 Labour–Fianna Fáil administration, who had been seen by police in an area of the city frequented by gay men, became the subject of enormous media attention when this fact was leaked to the media (by whom was never established). After tiptoeing around it for a number of weeks, the media eventually broke the story: Stagg confronted the issue directly in an RTÉ television interview, and retained his position. Later, the private life of Bertie Ahern, a senior member of Fianna Fáil who became taoiseach in 1997, also became a media issue. It was well known that he had separated from his wife of many years and was now living with another partner, but media generally left the subject alone until after a cryptic reference had been made to it by Albert Reynolds, when Reynolds was defending his leadership of Fianna Fáil against a challenge from Ahern.

Within the NUJ, ethical and professional issues, which had not generally been the subject of much discussion, suddenly achieved prominence in the early 1980s with the emergence of deep divisions on the question of abortion, which became a matter of intense political controversy – and eventually led to an amendment to the Irish constitution in 1983 (O'Sullivan, 1984). The problem had been incubating for some time. In 1975, the NUJ's annual delegate meeting (ADM) had pledged to agitate and organize to achieve the aims of a document called 'The Working Women's Charter', which included a policy for free and readily available abortion. The following year the ADM instructed the union's national executive council (NEC) to give active support to a pro-abortion-rights policy. By 1979, ADM delegates were criticizing the NEC for apparent foot-dragging on the issue, and the agenda for the 1980 ADM, which was held in April in Portrush, Co. Derry, featured a number of similar resolutions.

In 1980 the NUJ held a consultative conference on abortion in Dublin, attended by twenty-three women and fifty men (including a priest), which passed a resolution advising the NEC that union policy arising out of British legislation on abortion should not be extended to the Republic. A motion which, while accepting that the union did not need to involve itself in active campaigning on the issue, endorsed the right of women to choose an abortion was defeated by almost five to one. The controversy was defused, and the threatened major split in the Irish NUJ organization never materialized. There was, however, some ill-feeling: when one provincial journalist, Billy Quirke, left the NUJ to join the Institute of Journalists, he was 'blacked' under the terms of his paper's closed shop. He took the NUJ to court and secured payment of damages in an out-of-court settlement.

One of the by-products of these events was a growing feeling on the part of Irish NUJ members that their area of organizational and policy autonomy should be extended. This was eventually given practical expression, after a lengthy internal union debate, in 1993, when the Irish area council was replaced by an Irish executive council, with greater local autonomy over a wide range of issues. Another, less welcome outcome was that when abortion became a live issue in Irish politics in 1981–2, with the emergence of groups campaigning for and against the insertion of anti-abortion-rights clauses in the Irish constitution, journalists' attitudes, and NUJ policy, again came under the spotlight. The NUJ's ADM in 1981 passed a resolution supporting the Irish Women's Right to Choose Campaign, and instructing the NEC to support liberalization of legislation wherever the union had members. The Irish area council of the union distanced itself from this stance, but this aspect of NUJ policy, particularly during the national constitutional referendum on the abortion issue in September 1983, led to accusations by some supporters of the amendment that NUJ members involved in reporting the controversy were biased.

The evolution of broadcasting in the Republic during the period covered by this chapter was marked by a new Broadcasting Act, the first for a dozen years; by intensified controversies about the role of broadcasting in the Northern Ireland conflict (an issue that also affected Northern Ireland broadcasters, although in slightly different ways); by the organic growth of pirate radio, which had up to this point manifested itself only in a sporadic or limited way; and by arguments and decisions about the provision of programming choice in television.

The minister for posts and telegraphs, Conor Cruise O'Brien, had two major items on his agenda. One was the question of providing programme choice: in many areas of the country, television viewers could receive only RTÉ, whereas in the east and North, viewers had access to multiple channels due to the overspill from BBC and ITV. The second was the question of the responsibilities of RTÉ vis-á-vis the Northern conflict.

The previous government had already begun to address the matter by setting up in 1971, ten years after the passage of the original Broadcasting Act, the Broadcasting Review Committee. In February 1973, just before the change of government, that committee issued an interim report, rejecting suggestions that greater programme choice could be achieved by rebroadcasting either a BBC or an ITV service across the country, and recommending that a second RTÉ-controlled channel should be established that would carry, in addition to domestic programming, a wide range of BBC, ITV and other foreign programmes. O'Brien's preference, however, was for a different solution: the rebroadcasting of BBC1 throughout the Republic, coupled with the extension of RTÉ1 broadcasts to cover Northern Ireland.

The context in which these alternatives were being considered was significant. The 1973 RTÉ annual report (1973: 7) indicated that 77 per cent of the households in the state, or a total of 542,000 homes, had television sets. But only 530,000 of them had licences, and only 27,000 of them were in colour. The licence fee, at £7.50 (increased to £9 in October 1973) compared unfavourably, from RTÉ's perspective, with Sweden's (£19) and Germany's (£13): this helped to explain the fact that licence fees accounted only for 41 per cent of RTÉ revenue, compared with 80 per cent in Finland and 73 per cent in the Netherlands. Even these statistics, however, paled into insignificance beside the fact that the national station was attempting to provide a quality broadcasting service in the shadow of two of the best broadcasting networks in the world – BBC and ITV – whose overspill established difficult benchmarks for the indigenous service to meet.

A national debate ensued, culminating in 1975 with a survey which, when it was published in October, showed that a clear majority of respondents in both single and multichannel areas favoured the RTÉ2 option over the rebroadcasting of BBC1. O'Brien accepted these results, and gave permission to the authority to proceed on this basis, although, for financial reasons, the service was not to be inaugurated until 1978. Virtually simultaneously, strategic thinking within RTÉ was taking the reality of competition with increasing seriousness. RTÉ programmers decided early in 1977 to devise a programming policy to concentrate its resources to best effect in the prime-time slot, from 8–10 p.m. each evening (later extended to 7:30–10:30 p.m.). This policy 'was devised both in terms of our competitive role on the one hand and public service obligations on the other' (Mac Conghail, 1979: 10). In one form or another, this has been the backbone of programming policy ever since.

Another significant change had already taken place. The limit on the number of households that could be connected to a single high-specification television aerial, which had been set at ten in 1961, increased under political pressure to cater for a large working-class housing estate built on the outskirts of Dublin in 1966, increased again to 500 in 1970, and was abolished in 1974. The compensation for RTÉ was that aerial contractors would pay a percentage of gross rental income to RTÉ as compensation for the national station's presumed loss of advertising income (NAI, Taoiseach: TV 6027). Over the following 20 years, a process of rationalization took place, in the course of which RTÉ was to acquire ownership of Cablelink, a company that controlled the major part of the cable system, and that it eventually sold in two tranches in the late 1990s for more than £200 million.

These organizational and strategic choices, however, were to a considerable extent overshadowed by the arguments surrounding the new government's broadcasting bill, which was introduced in 1975 but which did not become law until a year later. O'Brien, whose suspicions about journalists in the print media have already been noted, was no less concerned and about the attitudes of journalists in RTÉ, for which he had, of course, more direct responsibility.

Because of O'Brien's known attitudes towards the IRA, frequently repeated and considerably hardened in the years after he left office, the received wisdom about the 1976 act is that its editorial interference in the way in which the Northern conflict was to be reported represented an unacceptable and unwelcome extension of the censorship imposed on the broadcast media by the previous government. It could also be argued, however, that, taken in conjunction with the fresh set of directives issued by O'Brien in relation to reporting political violence in Northern Ireland, it was in some respects a less restrictive, if not exactly liberalizing, measure. Still, it left in place a relatively authoritarian system for handling a complex series of political and journalistic problems, and this can be clearly seen in relation to the situation in Northern Ireland itself, where broadcasters continued to work without statutory restrictions on their professional activity until 1988 (Savage, 2015). This did not mean, of course, that they were unfettered: the system of internal checks and balances introduced by the BBC, and UTV's caution in relation to engaging with the problems at a serious level, gave rise to considerable, if intermittent controversy. But the contrast between the broadcasting dispensations in the two parts of the island was a very public illustration of the harsher regime obtaining for broadcasters in the Republic.

The 1976 act was based in part on the 1974 final report from the Broadcasting Review Committee, but rejected the committee's recommendation for the establishment of a broadcasting commission to supervise the whole broadcasting operation – an idea that was to resurface in various forms before finally being implemented as government policy in 1999.

In a number of areas the 1976 act represented a relaxation of the tight-fitting garment created by the 1960 act. It maintained the requirement for objectivity and impartiality in matters of current controversy, for example, but made it clear that balance could be achieved over a number of programmes rather than within the boundaries of each presentation. This was a common-sense solution that was already on the way to being adopted in practice. It contained a number of exhortations, such as those referring to the authority's duty to 'be mindful of the need for understanding and peace within the whole island of Ireland' and to foster awareness and understanding of 'the values and traditions of countries other than the State, including in particular those of such countries which are members of the European Economic Community' (Fisher, 1978: 43). The latter injunction, of course, also included Britain.

In other sections, the act, while retaining the right of the minister to dismiss the authority, laid down that he could not do so without first having secured majority approval in both houses of parliament for his decision; it created a mechanism whereby a ministerial directive could be rejected by either house of parliament; and it allowed the authority to depart from the general obligation of impartiality if it wanted to express its own opinion on matters of controversy connected with broadcasting policy. It established a broadcasting complaints commission (although the methodologies of the commission were so Byzantine that few disgruntled viewers had the stamina to carry their complaints to a conclusion); it instituted a system under which the standards of advertising would also be scrutinized; and it instructed the authority to avoid unreasonable invasions of personal privacy by programme-makers. It did not, as has been seen, make any provision for a second channel: this was dealt with separately by the minister.

It retained, as might be expected, the power of the minister to issue direc-tives to the authority. Even before the passage of the act, however, O'Brien had issued – in 1974 – his own directive, considerably more detailed than those that had been issued by his predecessors in Fianna Fáil. The detail it contained was evidence of the mounting frustration at RTÉ, where the authority's repeated requests to the government for clarification of the original directive had been met by the unhelpful assertion that it (i.e. the directive) spoke for itself. O'Brien's version, drafted in the light of numerous representations from RTÉ, went further in terms of providing detail. It not only specified the organizations whose members were to be banned from the airwaves, but inserted an open-ended provision ensuring that organizations proscribed by the British government would also be automatically included in the list of undesirables.

Even with these changes, the basic system remained unaltered: there was a level of direct and effective government editorial control of broadcasting which, despite intermittent protests by the NUJ and others, was to remain in place for almost another two decades. The Irish headquarters of the union in Dublin organized a questionnaire for its members in 1977 that evoked the comment, from one of them, that:

> The ultra-cautious atmosphere which Section 31 and the guidelines have fostered in the newsroom and programme sections has meant that enquiries into controversial areas have not been encouraged ... there is now a general anxiety about tackling stories which might embarrass the government on the issue of security. (Curtis, 1984: 194)

The situation was complicated by the fact that, within RTÉ, not all editorial personnel were of one mind on the issue, and there were strong anti-IRA sentiments, especially in the programmes (i.e. current-affairs) division, as opposed to the newsroom, where there would have been more sympathy for the republican cause in general, although not for the specific violence of the IRA. This reflected the division of opinion within the population generally: a 1979 survey by the Economic and Social Research Institute found that, while 41 per cent of the population of the Republic sympathized with the IRA's motives (a proportion closely resembling the 38–43 per cent who generally supported the Fianna Fáil party at elections), only 21 per cent actually supported their activities (Davis and Sinnott, 1979: 99).

The detailed clarification in the 1974 and subsequent directives of what the government actually had in mind prompted a revision of the guidelines issued by RTÉ to its staff, but was also to spawn further anomalies. In 1980, for example, RTÉ carried an interview with a member of the Irish Republican Socialist Party, the political wing of a paramilitary organization known as the Irish National Liberation Army. The INLA had been proscribed by British legislation, but the IRSP had not – whereas the IRA and its political wing, Sinn Féin, were proscribed by both governments. In 1981, RTÉ had to secure a special exemption from the government in order to show two episodes of the documentary series, *The Troubles*, because it contained interviews with people who were members of proscribed organizations (it was an RTÉ/BBC co-production, which underlined the absurdity of the situation). RTÉ, for its part, interpreted the directives with considerable rigidity, excluding from the airwaves people who happened to be members of Sinn Féin, even when their proposed participation in a programme was for reasons quite unrelated to their political role. A new situation arose in relation to a general election in February 1982, when Sinn Féin, by then a registered political party in the Republic, approached RTÉ to discuss whether the party would be entitled to party political broadcasts. The authority decided that democratic considerations should prevail, and announced that Sinn Féin would be facilitated in the same way as other parties. A major controversy ensued: the government issued a fresh directive prohibiting such Sinn Féin broadcasts; Sinn Féin appealed to the courts on the issue, without success.

As the newspapers were not bound by any of the restrictions that applied to RTÉ, it would be inaccurate to say that the public was deprived of all the information it needed. The consequences of the broadcasting ban, however, were pervasive and long-term. Although the immediate effect was to force broadcasters into a number of stratagems to make sure that viewers and listeners were not kept entirely in the dark (MacDermott, 1995: 70–80), the overall picture available to the public from the electronic media was necessarily an inadequate one. Long-term, the consequences included a yawning gap in the electronic media archive for more than twenty of the most politically significant years in twentieth-century Ireland, and a consistent tendency (in which, as it happened, the broadcast media were not alone) to underestimate the strength of Sinn Féin at election times. The existence of a blind spot is nowhere more tellingly illustrated than in McLoone and McMahon's *Television and Irish society* (1984), a study of twenty-one years of Irish television, whose otherwise cogently presented and well-researched essays managed to ignore Northern Ireland and the Section 31 directives completely.

Beyond politics, television was increasingly becoming a battleground for other interests as well. Irish-language groups, although to some extent mollified by the earlier establishment of Raidió na Gaeltachta, remained conscious of the relatively greater attraction of television as a popular medium and maintained pressure on the RTÉ Authority to increase programming in this area. In this area, inevitably, targets were more frequently set than achieved, although in 1976 television programming in Irish amounted to 10 per cent of total domestic programme output. This was a relatively high figure: Irish-language programming, as a proportion of total output and even of domestic production, suffered thereafter both from the extension of broadcasting hours and the introduction of RTÉ 2 in 1978. Women, too, added their voices to the clamour for more balanced representation. RTÉ set up an internal working party on the issue in 1979, a year in which its yearbook indicated, for example, that in the news division, all of twenty-four programme editors and presenters were men, and all but two of the seventy-two production assistants were women. This – and the fact that women were conspicuous by their absence from agricultural programmes – were among the issues raised at Dáil committee hearings.

There was also concern about perceived lapses from moral standards. Groups such as the Knights of St Columbanus and the St Thomas More Society complained about the inadequacy of the complaints procedure established under the 1976 act. Similarly, the Irish Family League alleged that RTÉ was '99 per cent weighted in favour of contraceptives, divorce, abortion, homosexuality etc.'. It also made a number of more specific criticisms, including one about a programme showing the 'full frontal' delivery of a baby, and a *Late Late Show* episode on which a woman had examined herself for breast cancer, which it described as 'indecent' (JCSSB: 111–13).

While RTÉ was defending itself against criticisms of these kinds, however, more serious problems of a structural nature were emerging. These were the challenge to its monopoly embodied in the emergence and rapid growth of pirate radio stations, and the resulting legal quagmire, which defied the sporadic and somewhat half-hearted government attempts to regulate them until the late 1980s. Until the early 1960s, pirate broadcasting had been short-term, and generally political in character. The department of justice maintained files on illegal broadcasts by republican activists in the 1940s and 1950s (NAI, Justice: S 164/40; S 16/57), but these broadcasts were generally short-term and related largely to events in Northern Ireland. In January 1959, at the height of the 1956–62 IRA campaign against Northern Ireland police and military installations, a broadcast from a pirate radio station describing itself as 'Irish Freedom Radio' was heard in Monaghan, on which an unidentified male speaker criticized the introduction of internment without trial by the Northern Ireland authorities (*IT,* 20 Jan. 1959). Such experiments were short-lived, not least because they generally attracted the keen attention of the security forces.

The emergence of pirate radio devoted to popular music, on the other hand, posed an entirely different set of problems and political imperatives. Irish government policy in this area was circumscribed for many years by two separate factors. One was the antiquity of the legislation (effectively, the Wireless Telegraphy Act of 1926) and the inadequacy of the penalties it prescribed. Successful prosecutions demanded proof that the accused possessed a radio transmitter; pirate radio operators, by removing a small component even as a police raid on their premises began, could successfully plead that their equipment was incapable of transmitting anything. The second was the popularity of the pirates: neither the UK nor the Irish government, having failed to move swiftly against pirate radio in the early years, was in a position to close down the stations without incurring serious popular displeasure. Stations like Radio Caroline (established in March 1964 by an Irishman, as it happened) created a new problem for both administrations, and led to a co-ordination of legislative action in a largely unsuccessful attempt to meet this new challenge (NAI, Taoiseach: S 17618/95; 97/6/169).

Initially it seemed as if the governments were winning the war against the pirates: the UK's Radio Caroline was put off the air in 1967, but reappeared in 1972. In Ireland, the pirates operated on a much smaller scale, and initially attracted little attention from the authorities. One of the earliest stations, known variously as 'The Colleen Home Service', 'Radio Laxey' and the 'Ballymun Home Service', operated intermittently from a North Dublin housing estate from 1944 onwards, mutating into 'Radio Galaxy' and 'Radio na Saoirse' ('Freedom Radio') in the 1960s and 1970s (Mulryan, 1988: 22–3). From the mid-'60s to the mid-'70s, however, pirate radio became a much more noticeable phenomenon, and attracted a considerable amount of police attention. The first successful prosecution of a Dublin pirate station,

Radio Melinda, in December 1972, attracted widespread publicity, not least because of the derisory penalties imposed. By the late 1970s, there were approximately twenty-five pirate stations operating from various locations, generally in the larger urban areas, attracting a listenership from young people tired of traditional radio, and annoyed at the heavy-handed attempts at suppression.

RTÉ, which had been concerned about the loss of advertising to these stations as well as by the implied threat to its monopoly, moved to defend its position, albeit somewhat lethargically. Its first community-radio experiment, Radio Liberties, was broadcast from a mobile studio in a working-class area of Dublin in 1975; similar experiments were carried out in Waterford, Limerick and a number of rural communities in that year and the next. It had a decentralized service operating from a studio in Cork, but of a very limited kind.

By 1978, RTÉ's response had moved up a gear. It was a two-pronged strategy. One prong was the launch of Radio 2, a pop-music station that would, it hoped, wean the listening public away from the pirates (and give the government a window of opportunity for moving decisively against them). RTÉ Radio 2 came on the air in a rush at the end of May 1979, and almost immediately captured 56 per cent of the listenership aged between 15–20 (Mulryan, 1988: 61). This figure, although hailed as a success by RTÉ, was also capable of a less flattering interpretation: a new station backed by all RTÉ resources had managed to attract only just over half of the potential audience. From this point up to and after the ultimate legalization of commercial radio, the battle for audience-share was relentless.

The second prong of RTÉ's strategy was a detailed set of proposals for a national community-radio service, approved by the RTÉ Authority at its meeting on 14 December 1979 and submitted to the minister for posts and telegraphs early the following year. These proposals expressed the authority's anxiety with some clarity:

> Any loss of advertising revenue to RTÉ would have serious effects on RTÉ's ability to serve the whole Irish community. Commercial radio is viable only in large cities but can, nevertheless, cream off a very important part of broadcasting income. This would seriously undermine RTÉ's capacity to provide services in other parts of the country and to be truly public in character. (RTÉ Authority minutes, 14 Dec. 1979)

These proposals argued that local radio could not be self-supporting in communities of fewer than 100,000–150,000 people, but suggested that with appropriate infrastructure a local radio could be created that would allow access on a regular (although unquantified) basis to every community of about 3,000 people. The initial phase of the service, they maintained, could be provided with a 3 per cent increase in the licence fee.

These proposals provided RTÉ, its trade unions, and some politicians with a standard around which to rally, but they were rapidly overtaken by events, and undermined by government inaction. Two government measures were drafted and introduced in 1980 and 1981 to take the pirates off the air and establish a mechanism for overseeing the development of legitimate commercial radio, but they lapsed with the onset of the 1981 general election. This failure coincided with the arrival of Radio Sunshine and, later, Radio Nova. These, described by the historian of this phase of Irish radio, Peter Mulryan, as the 'super-pirates' (1988: 84), were high-powered stations operating openly in defiance of the law, sometimes even carrying advertising from state bodies, paying taxes to the government, and flourishing in a legal and political vacuum. That vacuum was created in part by a lengthy period of political instability, which saw three general elections in eighteen months. In the run-up to the second of these, in February 1982, the country's largest political party, Fianna Fáil, took the unprecedented step of advertising its support for pirate stations in the national print media. Pirate radio responded by assisting the party to set up its own radio station at party headquarters (Mulryan, 1988: 93). Fianna Fáil won the election, but did nothing to legalize the pirates before they lost office in the second election of that year, in November. In the run-up to that election, the NUJ had retaliated, threatening not to interview on RTÉ any politician who broadcast on any of the pirate radio stations.

The new government comprised of parties opposed to Fianna Fáil took office determined to do something about commercial radio, but even as pressure to regularize the situation increased (apparently also from Britain, where the Irish super-pirate broadcasts could be heard with increasing clarity), the government found itself hamstrung by internal disagreement on the issue. Labour, the junior coalition partner, was steadfastly behind the RTÉ plan; Fine Gael, the senior partner, was broadly in favour of legalizing and regulating what could not be abolished, so that the government could move decisively against the more disreputable stations and in this way tidy up the airwaves. RTÉ's annual report for 1983 did not mince its words: the present situation was 'anarchic', with over seventy illegal operators, and the authority was not only disappointed by the absence of a government response to its local-radio proposals, but was hampered by 'the indecision regarding local radio' (1984: 8).

Misjudging the moment, the authorities launched a new and particularly fierce set of raids against the pirates. But these stations now had huge audiences, and the raids triggered a series of demonstrations which served notice on the government that the political costs of suppressing them were incalculable. In an attempt to buy time, and in the vain hope that it might engineer a politically acceptable response, it appointed the Interim Radio Commission in 1985 to study the questions involved. By the time this government lost office in 1987, the problem was no closer to a solution and the pirates, despite the best efforts of the authorities to close them down, and repeated moves by RTÉ to block their signals, were virtually mainstream media.

The conflict in Northern Ireland had, even before the mid-1970s, brought about as acute a degree of social and community polarization as had existed since the 1920s. The 1968–70 period, when it looked for a time as if the Northern Ireland Civil Rights Association would provide common ground for nationalists and unionists who wanted to bring about political reform without necessarily changing the constitutional status quo, was only a memory. The re-emergence of the Irish Republican Army, the more sporadic but equally lethal activities of loyalist paramilitary groups, and the increasingly controversial role of the British army, created a bunker mentality that reinforced traditional loyalties, especially in the print media. In 1973, a political initiative by the British and Irish and Northern Ireland governments known as the Sunningdale Agreement (after the Home Counties location where the talks took place) attempted to provide a new basis for stability, but succeeded only in making things worse. It was followed by a politically motivated strike in 1974 in which loyalist paramilitary forces operated to paralyse the Northern Ireland economy and, immediately afterwards, by the definitive imposition of direct rule from Westminster. If editorial polarisation had not been accentuated by the politics of the situation, the economics of what was happening would have encouraged it in any case. The Northern Ireland economy was in severe difficulties: part of the IRA policy was to wreck the industrial base of the province insofar as this could be achieved by paramilitary methods, and bomb attacks on British and foreign-owned business contributed to a situation in which advertising, in common with many other elements of economic activity like tourism, was severely affected.

There were a few exceptions. The *Belfast Telegraph* continued to draw readership and to some extent advertising from both communities; its politics remained resolutely unionist. The Northern Ireland edition of the *Sunday World*, although it was well known to be effectively a Dublin paper, maintained its position by de-emphasizing politics and sticking to its tabloid last. The major anomaly was the *Sunday News*, which, although it was a product of the unionist Century group, publishers of the *News Letter*, managed not only to combine Protestants and Catholics on its workforce, but to establish a measurable circulation in both communities.

> In the early 1970s, with riots, gunfire, 'action' up and down Belfast city, the paper had editions for the upper, mid and lower Falls [the nationalist area], the upper, mid and lower Shankill [the Loyalist heartland] … And research done on the readership constantly showed the breakdown spot on to Northern Ireland's Protestant/Catholic 60/40 split. (Barclay, 1993)

The launch of *Sunday Life* in 1988 by the *Belfast Telegraph*, however, brought considerable pressure to bear on the ailing *News Letter* company, and the *Sunday News* eventually expired in March 1993.

Some newspaper presses, notably those of the Mirror group outside Belfast, and of the *Belfast Telegraph*, were the intermittent objects of the violence. After a 1976 attack on the *Telegraph's* presses a four-page 'penny marvel' version of the paper was produced on printing presses owned by a local newspaper just outside Belfast, and sold some 75,000 copies a day until normal production was resumed (Brodie, 1995: 180–2).

A 1984 survey showed that the *Telegraph* was being read by 40 per cent of the adult population of Northern Ireland; the figure would undoubtedly have been higher but for the polarization of opinion, which was weakening its appeal to nationalist readers. Under Thomson's stewardship, however, it continued to invest for a future it no doubt hoped would arrive sooner rather than later: in April 1985 it commissioned its first colour presses.

The same year saw a political development that some hoped would finally reduce, if not eliminate, violence in Northern Ireland. This was the 1985 Hillsborough Agreement between the British government led by Mrs Thatcher and the Irish government led by Dr Garret FitzGerald. Although crafted with more care than its 1973 predecessor, it generated similar suspicion and antagonism on the wilder shores of both unionism and nationalism, suspicion that the *Telegraph* shared, to a degree. 'The price of nationalist consent to participation in Northern Ireland, if this it is', it editorialized the following day, 'has been fixed at an unrealistically high level' (16 Nov. 1985).

Its contemporary, if not exactly its rival, the *Irish News*, saw things in a predictably different light. Hillsborough, it said, was 'a brave and commendable attempt to begin the healing process' (*IN*, 16 Nov. 1985). But by then the *Irish News* itself had already begun a process of change. In 1982, it had acquired a new editor in Martin O'Brien, a 27-year-old university graduate who had served his apprenticeship on the *Belfast Telegraph*. Even as the conflict worsened, O'Brien came to the conclusion that his newspaper would have to take a stand, however limited, against the glorification of political violence by some elements within the nationalist community, or at least against his paper's apparently ambiguous attitude to the IRA. He did this by putting an end to the type of death notice normally inserted in the *Irish News* on behalf of members of the IRA who had been killed in the conflict by members of the security forces. Two years later, he was to leave for a position with the BBC, but the changes he had inaugurated had taken root, even as the paper's ownership was changing. In 1983, half-way through his period as editor, the two members of the McSparran family who were the principal controllers of the newspaper died in an accident, and the paper came into the hands of another Northern Catholic business family, the Fitzpatricks.

The print media were to some extent isolated from the intensity of the conflict by the fact that their partisan coloration was, by and large, unmistakeable. The electronic media were in a different position, as they were established and regulated by the state on both sides of the border. In the Republic, the successive directives issued by governments under the Broadcasting Acts had created a sort of eerie silence, in that the voices of some of the principal protagonists had been stilled; in Northern Ireland, on the other hand, the absence of such a clear policy created a no man's land for broadcasters in which skirmishing was the order of the day. Whatever the broadcasters did, they were closely watched:

> Roman Catholics and Protestants alike are avid news-watchers. They could never be accused of indifference to the image of themselves that appears almost nightly on their television screens. Often, I believe, they switch channels keenly in pursuit of it. They share especially in

television's immediacy – being often the witnesses of an event itself and shortly afterwards of television's account of it ... In sectarian conflict, it is not uncommon for each side to regard the camera as the instrument of the other's propaganda and to react to it accordingly. (Bell, 1973: 371)

The guidelines that had been established by the BBC and by ITV in 1971 remained the operative ground rules for the reporting of the conflict, but they did not succeed in averting controversy (Savage, 2015). They were supplemented, although not directly, by two pieces of Westminster legislation. The first, the 1973 Broadcasting Act, forbade any incitement to crime or disorder, and this provision was explicitly cited in a decision by the Independent Broadcasting Authority to ban a programme in 1977 that featured an account of a speech in Dublin by a member of Sinn Féin, even though the recording of his actual words had been excised (Curtis, 1984: 155). The second was the 1976 Prevention of Terrorism Act. Although it was suggested during its passage through parliament that it should be amended to specifically prohibit broadcasters from aiding or abetting terrorists, no amendments of this nature were made, on the understanding that such would not be necessary. In the view of a commentator with a relatively benign view of the activities of the BBC management during this period, the effect of this particular piece of legislation was:

> almost certainly equivalent to the directive which the Fianna Fáil government issued to the Irish broadcasting service, Radio Telefís Eireann, forbidding the service to broadcast interviews with subversives or film of their activities. (Cathcart, 1984: 240)

When this comment was written, no interviews with paramilitaries and almost no film of their activities had been transmitted by the BBC since 1981. This sanitized coverage had implications for elections. The 1982 elections to the Northern Ireland Assembly, a short-lived body designed to take the place of the old Stormont parliament, saw the election of Gerry Adams to one of the Belfast seats with almost 10,000 votes, but coverage of this aspect of the election was severely limited by the legal prohibitions affecting broadcasters generally, particularly in the Republic.

Nor had the situation been much more open prior to that, since the overall tension between broadcasters and government had been heightened by an incident in Carrickmore, Co. Tyrone, in 1979, when a BBC unit had filmed

IRA men mounting an illegal roadblock. This had led directly to a tightening
of the 1971 guidelines in both broadcasting organizations. Any list of controversial broadcasts involving ITV would of necessity include a number of
interviews carried out by the British journalist, Mary Holland, with republican leaders for that network in 1973 and 1974. In the case of her 1974
interview, the tension was heightened by the fact that, four days after it was
broadcast, the Birmingham IRA bombings, in which 21 people died, took
place. This led some political commentators to argue that there was a direct
link between the interview and the explosions (Curtis, 1984: 160). The IBA
was also involved in controversies surrounding the publication of a report by
Amnesty International in June 1978. A planned programme on this report was
banned by the IBA, an action the reporter involved, Peter Taylor, described
bluntly as 'political censorship' (Curtis, 1984: 62). Similar difficulties attended
the broadcasting of some programmes in the series *The Troubles* in 1981. The
role of the domestic contractor, UTV, in all of this was the subject of a
veiled, and somewhat tortured contemporary remark by UTV's chairman,
R.B. Henderson, who commented that:

> on many occasions, particularly in the last decade or so, I had some
> difficult decisions to take on programmes made by 'visiting firemen' on
> Northern Ireland. We were mindful of their effect not only in Northern
> Ireland and upon our viewers but also upon national views and even
> international ones. (Henderson, 1984: 43–4)

If the difficulties experienced by the IBA in relation to the Northern Ireland
conflict seemed to be related primarily to current-affairs coverage in the
general sense, those at the BBC extended much more deeply into the more
limited, but politically more sensitive news area. News always had a greater
potential for conflict, but had been much more cautiously handled by both
ITV's national network and by its Northern Ireland franchisee, UTV. This
problem surfaced dramatically in May 1974, during the Ulster Workers'
Council national strike which successfully aimed at bringing down the power-
sharing executive in Northern Ireland created as the outcome of the
Sunningdale Agreement. The UWC had prepared plans for a pirate radio
station of its own, but never went ahead with it because, in the words of one
UWC leader, the BBC 'were prepared to be fed any information. They fell
into their own trap that "the public must get the news"' (Cathcart, 1984: 231).

The BBC was accused by the Northern Ireland secretary in 1976 of providing 'a daily platform' to the IRA (Cathcart, 1984: 234), and hostilities deepened the following year when a programme by the veteran broadcaster Keith Kyle for the *Tonight* team raised serious questions about the British government's commitment to end the inhuman and degrading treatment of persons in police custody, which had been identified in 1971 and had been condemned by the European Court of Human Rights at Strasbourg.

Worse was to come when, with the IRA hunger strikes of 1980 and 1981, the BBC found itself in crossfire more intense than any it had yet experienced. Northern Ireland nationalists were incensed by what they saw as pro-government reporting; unionist leaders – one of whom described a *Panorama* programme in September 1981 as 'a propaganda coup for the IRA' (Cathcart, 1984: 245) – saw the whole question from a diametrically opposed perspective.

On its way to achieving a stance that might, in the eyes of its defenders, be described as offending everyone more or less equally, the BBC nonetheless came in for some criticism. Martin Bell, who in 1972 had defended his employers against charges of censorship, still maintained in 1979 that 1969 had been the only occasion on which the BBC had engaged in unacceptable self-censorship. This limitation of unacceptable self-censorship to the 1969 period was noted critically by some writers such as Hall (1973), Curtis (1984), Woodman (1984) and to a lesser extent Butler (1996). Harshest of all was Raymond Williams, who commented:

> The BBC has in general, and especially in recent months, a less than honourable record on the reporting of events in Northern Ireland. The point has been reached ... when some views are excluded because they are *prima facie* abhorrent – the charter of censors anywhere in the world. (1973: 15)

The situation was to change yet again in the late 1980s, when further, this time statutory, restrictions were introduced on the broadcast media, but the Northern arena had by then become the scene for increasingly sophisticated attempts at news management by all involved. The introduction of direct rule in 1974 and the extension of the activities of the Northern Ireland Office in the years that followed led to an intensification of media-management techniques. The involvement of the British army, which had its own press operations (which included disinformation activities), supplemented the already well-developed RUC press office. In 1990, the Northern Ireland

Information Service spent £7.2 million in the North for a population of 1.5 million, on top of the considerable sums expended in media management by the police and army (Feeney, 1997: 44). Sinn Féin, too, developed an increasingly agile and focused press-relations strategy during this period.

The smoke from the battleground of news and current affairs appeared to smother creativity in other fields also. At a conference in London in 1980, dramatists and critics complained that Northern Ireland had become off limits as a subject for television drama, and that only nineteen television plays had been produced in twelve years on the themes of that conflict (McLoone, 1996: 80). It was certainly notable that, even before the conflict erupted in 1968, the few television dramas that actually dealt with such topics had generally been centrally produced by the networks, rather than by the – admittedly small – regional centres in Belfast. These included Sam Thompson's *Over the Bridge* (Granada, 1961) and *Cemented with Love* (BBC, 1965); Alun Owen's *Progress to the Park* (BBC, 1965); and John D. Stewart's *Boatman, Do Not Tarry* (UTV, 1968).

7 Coming to terms, 1985–95

The fluidity and volatility that characterized Irish media from the end of the 1970s into the mid-1980s persisted and even intensified in the decade that followed. In the electronic media, the already qualified monopoly of the state was further fragmented as the airwaves were opened up to commercial interests in a profoundly significant restructuring of the legislative and regulatory framework. In print, a process of rationalization began to take place on both the national and regional level; new titles appeared to target niche markets as part of a move away from generalist media; and Irish media, both print and electronic, became the object of increasing interest on the part of non-national entrepreneurs and companies. In the process, the sharp line that could traditionally have been drawn between indigenous and non-Irish media became blurred, raising new issues of ownership, control and content.

The establishment of National Newspapers of Ireland in 1985 was especially significant in the circulation war being waged between indigenous Irish titles and the UK imports. It was a war in which, initially, it was difficult to discern the relative strengths of the combatants. Most figures for UK imports were, until the mid-1990s, no more than industry estimates, and Central Statistics Office figures for newspaper imports took no account of unsold or damaged papers.

A 1978 study had suggested that imports appealed to the 'younger and the less inhibited sections of the community' (M.H. Consultants, 1978: 12). It added:

> The view was expressed by more than one publisher that there was scope for only three Dublin dailies: one at the top of the market (the *Irish Times*), one in the middle (where there are currently two, the *Independent* and the *Press*) and one at the lower end of the market (where there are currently no Irish papers). (M.H. Consultants, 1978: 12)

This analysis was to be borne out, with eerie accuracy, in the period under review, but in ways in which the authors of that prescient report could barely have predicted. The *Irish Times*, which Douglas Gageby had returned to edit for the second time in 1986, had again recorded substantial circulation gains between then and the early 1990s (although it was still well behind the *Irish Independent*). The *Irish Press*, however, was faltering. In 1981 it had reached

104,000. Then its circulation began a remorseless slide, assisted by a strike in 1985 that kept it off the streets for twelve weeks. By 1988 it had fallen to just under 66,000, and not even the cross-subsidization from the still successful *Evening Press* and *Sunday Press* could act as a lifeline forever. The *Evening Press*, in fact, declined by more than 20,000 copies to less than 99,000 between 1987 and 1989. It was not alone: changing commuting habits, and competition from other media, including afternoon television, were responsible for a persistent decline in the circulation of both Dublin evening papers throughout the 1980s and early 1990s, despite the fact that they had no direct UK newspaper competitors.

The Independent group's role in the marketplace was coming in for increasing scrutiny. One international study (Sanchez-Taberno, 1993: 101) noted that in 1990 it controlled 51 per cent of the market, the second highest share controlled by a single company in any country studied, after Austria (where Mediaprint controlled 54.4 per cent). In Ireland, the same report noted, the market share of the two largest groups, the Independent and Press groups, was 75 per cent, a figure not exceeded by any other country (Austria was next, with 68 per cent; Switzerland was lowest with 21 per cent). The larger group, although outwardly successful, also had problems, particularly competition from the UK tabloids, which were nibbling away at the circulation base of its popular morning daily.

Both the Independent and the Press groups now moved to shore up their positions, with strategies that were superficially similar, but that had dramatically different outcomes. The mechanism chosen in both cases was co-financing with a foreign source of capital. The venture by the Independent group was, after a shaky start, to produce extraordinarily beneficial results; that of the Press group was to end in catastrophe, and the closure of all the group's newspapers in 1995.

Even before its financial reorganization, the *Irish Press* had moved to meet the tabloid threat in its own way. In an attempt at diversification, it had acquired *Southside*, one of the oldest and most successful suburban free sheets; but the market for free sheets was dwindling, and *Southside* was closed down in 1987. In the same year, Tim Pat Coogan, who had edited the *Irish Press* for almost two decades, was eased out. A new editor, Hugh Lambert, was installed, and UK consultants were employed to oversee the daily paper's transformation from a broadsheet into a tabloid, which took place early in 1988. This was not a success. Desperately short of capital, the management initiated negotiations on refinancing in December 1988 with a US newspaper magnate named Ralph Ingersoll, who had inherited a substantial network of newspapers from his father and was reputedly worth $200 million. These negotiations, when they concluded in November 1989, produced a commitment by Ingersoll to invest £5 million in the company, and the creation of a complicated network of companies, in some of which Ingersoll and the young Eamon de Valera, grandson of the founder, shared control, with 50 per cent each, and in others of which – notably the company owning the titles – de Valera retained personal control. In May 1990 a reorganization plan was agreed with the workforce.

Over at the Independent group, however, matters were moving more swiftly. For some time, its management had been in negotiations with United Newspapers in London, publishers not only of the broadsheet *Daily Express* but of the tabloid *Daily Star.* The latter paper sold some 47,000 copies in the Republic. These negotiations resulted in the creation of a jointly owned company – 50 per cent on each side – to publish an Irish edition of the *Star,* printed on the colour presses owned by the *Sunday World,* now part of the Independent group. The new *Star,* which was published for the first time on 29 February 1988, was at the cutting edge of newspaper technology in Ireland, in that its print generation was entirely computer-based, and involved a high-speed computer link to the Express group in London. Whole pages could be – and were – transmitted electronically to Ireland, where they were blended with domestically generated material to produce a new hybrid paper. Initially, the experiment did not seem to be a marked success. The formula adopted for the paper was too closely aligned with that of the UK imports, in that it printed relatively little Irish-generated news and features, and it lost money. After a redesign in 1990, the amount of Irish coverage increased dramatically, and the circulation figures increased. By 1992 its circulation was up to 85,000, which contributed to a substantial rebalancing of the circulation figures for Irish newspapers generally, as it was now classified as an Irish paper, whereas its predecessor, the *Daily Star,* had been classified as a UK import.

Its editorial differences with the UK *Daily Star* were quantitative and qualitative. It tended to use scandal unconnected with sex as a circulation-builder more frequently than the UK paper; its astrology and agony columns were different; its section for female readers was entitled 'Star Woman', compared with the UK paper's 'Star Girl'; and additional clothing – generally a bikini top – was judiciously added to its cartoon character Katie to obviate the possibility of protests from Irish readers. These changes, however, were probably less important in the long run than the pagination: one 1996 sample showed the *Star*, with 38 out of its 48 pages characterized by Irish content, substantially ahead of its tabloid rivals the Irish *Mirror* (11/36) and the Irish *Sun* (7/36). It began, even, to make inroads into Northern Ireland, not least because of its substantial coverage of Irish sports; it also employed Ireland's first female sports editor.

The technology available to the *Star* and the *Sunday World* was in sharp contrast to that available to the Press group for the production of its titles. Not only were the rotary presses old, and causing maintenance problems with increasing frequency, but they could not print colour as part of the normal printing operation, so that colour advertisements or photographs had to be pre-printed, with a necessarily long lead time. The *Irish Independent* managed to get a colour photograph of the Irish cyclist Stephen Roche onto its front page the day after he won the Tour de France in 1987; but keen-eyed observers noticed that it was a photograph taken earlier in the race.

Partly for this reason, and partly because the multinational media proprietors Rupert Murdoch and Conrad Black were rumoured to be planning printing facilities on the island, the rumour was floated in 1990 by Independent Newspapers that there was a possibility of a new consortium that would finance a purpose-built printing press to print not only the Independent titles, but the *Irish Times* and the *Cork Examiner*. This overture, if overture it was, evoked little interest from the other two managements, each of which had already re-equipped, and saw no need to help a rival solve its technological problems.

The group's need to refinance its physical plant did not, as it happened, deter it from further expansion. Its new ventures included diversified communications and media interests in Ireland, the UK, France, Mexico and Australia (where it owned a transit-advertising company and a 14.95 per cent share in Australian Provincial Newspapers). Simultaneously, it was building up a shareholding in the UK Mirror group, which was in financial disarray following the unexpected death of its controlling shareholder, Robert Maxwell, and in the UK Independent group, publishers of the *Independent* and *Independent on Sunday*. It had created a new company, Independent Wireless Cable, to spearhead its involvement in the television-signal distribution business. This was to resurface dramatically as a political issue involving the Independent group in 1997.

Although the *Washington Post*, where Dr O'Reilly, Independent Newspapers' chairman, was a board member, was projected in 1990 to purchase some 20 per cent of the group, this deal did not materialize. Had it taken place, this would have been the largest incursion into the domestic Irish media market of any foreign interest, and would have triggered alarm bells in many quarters. It did not happen, but it was not the last time that rumours connecting the Independent group and major foreign interests were in circulation (Truetzschler, 1991a: 2).

The Independent group in November 1990 turned its attention to the *Sunday Tribune*. The *Tribune* had been making a steady, if modest, profit since its recapitalization in 1986, up to 1989/90, when it made £245,000. Shortly afterwards, a downturn in the advertising market and related economic factors indicated that a bleak future was in store. Approaches were made to both the *Irish Times* and Ralph Ingersoll, but 'to little effect' (Competition Authority, 1992: 5.6). The alternative – a gamble by any standards – was the launch, in May 1990, of the *Dublin Tribune,* a multi-edition free sheet (it had no fewer than nine different editions), which outran its budgets in spectacular style, bringing the whole enterprise to the verge of collapse. The profit of £245,000 became, in a year, a loss of £2.3 million. This, and the fact that a large parcel of *Tribune* shares was known to be on the market, was the context in which Independent Newspapers acquired a 29.9 per cent shareholding in the paper, at a cost of £805,000. This transaction took place without the knowledge of the *Tribune*'s board or its editor, Vincent Browne, who stated that 'at first glance, at least, it seems to threaten the independence of the newspaper and the company' (*ST,* 18 Nov. 1990). The Independent group's rationale appears to have been twofold: apprehension that the shareholding on offer might become available to the *Irish Times,* which did not publish a Sunday title; and apprehension that the collapse of the *Tribune* would open the 'quality' Sunday market in Ireland up to the predations of, in particular, the *Sunday Times,* with equally deleterious consequences for the *Sunday Independent.* The Independent group's board's description of the purchase of *Tribune* shares as an investment (Competition Authority, 1992: 5.16) was difficult to reconcile with the *Tribune*'s financial performance.

The *Irish Press*' situation was going from bad to worse. It had relaunched its evening paper as a two-section publication in 1988, but the change succeeded only in alienating readers, and circulation slumped again. In the same year, its chief rival, the *Irish Independent,* launched an expensive bonus promotional game for readers called Fortuna, which 'almost wiped out' the *Press,* but which was stopped after the *Press* took court action against it (Competition Authority, 1995: 4.40). In 1990, more significantly, relationships between the de Valera directors and the Ingersoll directors began to sour. There were disagreements over appointments to key positions. One executive was dismissed by the Ingersoll side of the management, only to be rehired almost immediately by the de Valera side. Privately – it emerged later – Ingersoll was already looking for new investors.

In May 1993 the two sides of the company went to court, as they were by now totally unable to resolve their differences. Shortly afterwards, the de Valera side initiated brief and fruitless negotiations with the Daily Mirror group in an attempt to source replacement capital. The court case, which it had been thought would last for six weeks, dragged on until December, consuming much of what capital was left to the business. The court accepted the contention of the de Valera management team that Ingersoll had behaved improperly, and, as part of its judgment, required Ingersoll to sell his shares in the jointly owned companies back to the de Valera side, which of course involved a further drain on the company's resources.

The Irish Press group was by now plainly in a desperate state, and predators, or suitors, began to circle. Although its circulation was plummeting, it was not destitute: it owned valuable property in the centre of Dublin, and a substantial shareholding in the Press Association. It needed investment and – possibly – a new management team if it was to be relaunched and survive.

In the autumn of 1994, the first substantial offer materialized. This was from a consortium formed by the *Daily Telegraph* group in London, owned by the Canadian Conrad Black, and a new Irish Sunday paper, the *Sunday Business Post*. When it became clear that the replacement of the *Irish Press* management team was a distinct possibility, the proposal was rejected. A factor in the rejection was undoubtedly the approach by Independent Newspapers, which indicated that the company was prepared to give the Irish Press group a £2 million loan in return for a 24.9 per cent shareholding. No management changes were envisaged.

After the latter proposal had been accepted by the *Irish Press* management, the government asked the competition authority to examine the extension of the Independent group's role in the Irish newspaper market and report on issues of possible dominance and the implications of these for competition. The authority heard evidence from a wide range of interested parties, and came to the unambiguous conclusion in a 1995 report that the transaction represented 'an abuse of a dominant position by Independent Newspapers … and … an anti-competitive agreement' (8.62). It recommended that the minister should take a court action under the Competition Act 'seeking an injunction and or declaration prohibiting these arrangements' (8.64). As the minister, Richard Bruton, considered his options, the situation deteriorated rapidly. Colm Rapple, a respected *Irish Press* financial journalist, was dismissed by the company after he had published – in the *Irish Times* – an article about his own newspaper's finances. The NUJ retaliated by holding mandatory meetings of the Irish Press group journalists, effectively halting production of the titles on 25 May 1995. The title was never published again. The journalists produced a desktop newspaper, the *Irish X-Press*, at various dates between May and July, which helped to raise money for benevolent purposes and gave the dispute between management and the union a continuing high profile, but the titles never reappeared and the subsidiary companies in the group eventually went into liquidation, leaving the de Valera interests with some liquidity, based largely on the Press Association shareholding.

In the meantime, the Independent group, taking stock of its position in relation to the *Sunday Tribune*, realised that it was locked into a difficult situation. It had two nominees on the board, but the company's financial performance had not been turned around and, at the beginning of 1992 the free sheets, which had by then lost over £2 million, were given a warning that if they did not become profitable after 13 weeks, they would close. Accumulated debts were mounting, and profitability seemed as unattainable as ever. The *Sunday Tribune* itself continued to make headlines. Among other notable exclusive stories, its reporter Veronica Guerin secured an interview with the Catholic bishop, Dr Eamonn Casey, who had left the country after news broke that he had fathered a child and misappropriated diocesan funds. This achieved huge sales for the paper over a period of some weeks; but the additional revenue was, to a considerable extent, offset by heavy promotional costs. The Independent group now moved to assume control of an operation that it was supporting financially, and which was haemorrhaging funds, by proposing to increase its shareholding to 53.09 per cent. It was referred to the competition authority in February 1992, and was made the object of the first enquiry carried out by that authority under the 1991 Competition Act.

The authority's report recommended against the proposal, on the grounds that 'the proposed merger would be likely to prevent or restrict competition and to restrain trade in the market for Irish Sunday newspapers … and that it would be likely to operate against the common good' (Competition

Authority, 1992: 6.17). The Independent group now had to take some hard decisions. It had previously told the authority that:

> if the rights issue did not proceed, the Tribune would shortly go into liquidation. The Independent, having lost its investment of around £1m, would then have no further involvement in the company. (Competition Authority, 1992: 5.3)

However, for undisclosed reasons, it now decided that it had no option but to keep the *Tribune* afloat. Its cost/benefit calculations would have included at least the following factors: the threat to the *Sunday Independent* from a newly aggressive *Sunday Times*, which was increasing its Irish editorial and marketing budgets; the danger of leaving a gap in the 'quality' end of the Irish Sunday market that the new and hungry *Sunday Business Post* would be only too eager to fill; the risk of exposing the Independent group to further accusations that it now held almost total control of the indigenous Sunday newspaper market in Ireland, and possibly even to action under the Competition Act on this score; and the public relations debacle that would have ensued for Independent Newspapers.

These factors evidently weighed more heavily with the Independent board than the ongoing costs of subsidizing the *Tribune*. These were now so high that, effectively, regular cash transactions by way of loans from the Independent group were, from then on, the sole reason why the *Tribune* remained on the streets. The Independent group now had effective control of the *Tribune*, despite the rejection of its takeover strategy by the competition authority. This was most clearly demonstrated in 1994 when the board of the *Sunday Tribune*, which now included several representatives from Independent Newspapers, terminated Browne's appointment as editor at two hours' notice. He was succeeded, first, by Peter Murtagh, a former *Irish Times* journalist who had spent nine years with the *Guardian* in London, and then by Matt Cooper, a financial journalist from the staff of the *Irish Independent*. Under Cooper's editorship, the paper's finances improved slowly, but – although it was making money intermittently on an issue-by-issue basis towards the end of the 1990s – its accumulated debt to the Independent group by 2000 was of the order of £12 million.

This underlined the fact that what had occurred was, and remained, a messy solution. The Independent group was precluded from translating the *Tribune*'s indebtedness into equity, or offsetting its losses against group profits.

Equally, that indebtedness, and the Independent group's 29.9 per cent equity, operated as a major deterrent to any potential investor, Irish or foreign, who might otherwise have been interested in taking over the *Tribune* as a going concern. Existing Irish media groups, in particular, would not find it easy to make a foray into the Sunday newspaper market on terms that immediately refunded a large sum of money to the most powerful competitor in the marketplace. Finally, this continuing subsidy of the *Tribune* undermined the logic of one of the campaigns being waged by the Independent group and other NNI members against UK imports: a campaign against below-cost and promotional selling. Effectively, there was now no definition of below-cost selling that could be deployed in this campaign that did not also catch, *a fortiori*, the *Sunday Tribune.*

Despite the fact that the Independent group did not technically own the *Sunday Tribune*, its position in the Irish Sunday market was now quite extraordinarily powerful. It directly owned the two largest selling papers, the *Sunday Independent* and the *Sunday World*, which between them had a circulation at the end of 1991 of 477,500 (excluding the *Sunday World*'s circulation in Northern Ireland). Each of these papers, on its own, was outselling the *Sunday Press*' 200,500; and the *Sunday Business Post*, at 28,300, was still a minnow. If one counted the *Sunday Tribune*, the Independent group's share of the indigenous Irish Sunday market was 71.2 per cent. But UK papers – notably the *News of the World*, the *Sunday Mirror* and the *Sunday People* – with a combined Irish circulation of almost 300,000, still accounted for 32 per cent of the total Irish market, a factor repeatedly adduced by the Independent group to buttress its claim that, taking the market as a whole, its position was not a dominant one.

The existence of the *Sunday Business Post*, at the other end of the scale from the Independent giant, was an anomaly in itself, but one which – after a number of vicissitudes and some good luck – was to prove that tenacity, vision, and an appropriate publishing formula could still generate significant new opportunities in journalism and publishing in Ireland. It was the brainchild of four journalists and became, effectively, the first modern newspaper to be launched in Ireland by journalists rather than by businessmen or politicians. They were Damien Kiberd of the *Irish Press*, Frank FitzGibbon of the monthly *Irish Business*, Aileen O'Toole, then editor of *Business and Finance*, and James Morrissey of the *Irish Independent*. In 1989, when they brought their plans to fruition, no Irish newspaper, daily or Sunday, had a separate business and finance section. Their plans for a weekly financial broadsheet did not exclude a weekday publication, but they eventually settled on Sunday – despite the fact that this was where the competition from UK imports was fiercest – because this was the day of the week on which, traditionally, most newspapers were purchased.

Two Irish stockbroking companies and an Irish regional newspaper – the *Kilkenny People* – became investors in the project, and the new paper was to be printed in Kilkenny. But the financing arrangements fell apart twenty days before the first issue was due to appear, and the founders were left with a staff, computers and no funds. A French investor, Jean-Jacques Servan-Schreiber of Groupe Expansion, who had heard of the project and expressed an interest, stepped into the gap at short notice and took a 50 per cent share-holding. The launch, on 26 November 1989, was chaotic, as the original printing arrangements with the *Kilkenny People* had collapsed. The first three issues of the paper were printed by three different companies, before it eventually settled down with Drogheda Web Offset, in which company it later took a 25 per cent stake. By the end of its first year of publication, against expectations, its circulation was 26,000, considerably ahead of its target of 17,000.

It was in the process of articulating a formula that would prove increasingly popular with the business community. It saved considerably on editorial costs by taking a firm decision not to emulate other Sunday newspapers' extensive coverage of sport (although it did carry articles on financial and business aspects of sport). Politically, it resembled the ailing *Sunday Press* more than the *Sunday Independent* or *Sunday Tribune*, in that it hewed to a broad nationalist agenda. But it garnered circulation not only from its extensive coverage of business (often with hard-hitting business stories nestling uncomfortably beside soft-centred advertorial), but from a succession of journalistic exclusives generated by adventurous, often young and female, journalists. These included Veronica Guerin (O'Reilly, 1998), who began what was to be a spectacular career with the *Post* as a freelance in 1990 and specialized in writing about fraud and white-collar crime, and Emily O'Reilly, its political correspondent. Its design, by Stephen Ryan, won it Newspaper Design Awards in Britain and from the Society of Newspaper Design in each of the three years 1991–3.

Its teething problems, however, remained substantial. In June 1991 FitzGibbon left, after costs had begun to escalate: the original plan for a twenty-four-page paper had given way to a sixty-eight-page paper with a staff of sixty. In 1992 Morrissey left, but in the same year Barbara Nugent joined the company from the *Sunday Tribune*. She was to prove a valuable asset to the

Post, indicating the measure of her loss to the *Tribune*. At this time, tentative moves to establish a bridgehead, and even a possible merger, between the two publications, foundered, not least because of the strained relationship between Nugent and her former employer. In 1993, there were two further significant developments. The first was the move to Belfast, where the *Post* was now to be printed by the *Belfast Telegraph*. This had cost savings and technology advantages, although it necessitated an earlier deadline – initially Friday afternoon – which meant news that broke on Saturday could not be covered in the paper on sale the following day. In the long run, this was a problem that did not affect the *Post*'s readership or circulation to any significant degree. The second was the departure of Servan-Schreiber, who reduced his stake in the company to 10 per cent, and was then replaced by a German entrepreneur, Norman Rentrop. Even more critically, Servan-Schreiber wrote off his investment in the *Post*, which stood at more than £2.5 million. This effectively relaunched the paper on a considerably more secure financial footing. By March 1995, the Joint National Readership Survey figures showed that the *Post*'s readership had increased by 14 per cent to 105,000, and that it had the highest proportion of social-class AB readers of any national newspaper: 49.5 per cent, as against 25 per cent for the *Sunday Tribune* and 16 per cent for the *Sunday Independent*, its direct Sunday competitors; and 40.9 per cent for the *Irish Times*. This had important and enduring consequences for advertising, as well as for circulation, which was now at 33,000.

The *Irish Times*' circulation was also showing a steady increase in the decade after 1985, despite the fact that the economic boom, as such, did not really develop a full head of steam until 1995. A new colour press, which arrived in 1986, allowed the paper to have colour in editorial and in advertising, and to increase the size of the paper. In 1986, it was carrying an average of 108 columns per day; by 1996 that had grown to nearly 220. The average size of the paper in 1986 was eighteen pages daily; fifteen years later, it was in excess of fifty pages.

There were a number of other, related, editorial developments, notably increased sectionalization, aimed at particular groups. The paper also invested in staff. At a time when editorial numbers at the *Independent* were more or less constant at around 250 journalists for the three titles, the *Irish Times*, which had about 190 editorial staff in 1986 for its two titles (including the equine weekly *Irish Field*), had expanded ten years later to figure, including full-time contract staff, about 280. There was a conscious shift to go further into specialization: science, legal affairs, environment, health, social affairs, media and food. It also developed a number of overseas offices. In 1986, its furthest-flung office was Brussels. Ten years later it had offices in Moscow, Washington, Beijing, Paris, Rome and Bonn, and a lot of new stringers. Then in 1996 it further expanded its regional coverage.

Part of this was due to the *Irish Times'* unusual corporate structure. It was making comfortable profits, but had no shareholders, and therefore did not have to pay dividends. This left more finance available for expansion, particularly on the editorial side. On the other hand, it had – for the same reason – no way of raising additional capital for re-equipment and new presses, short of borrowing, so had to maintain a large liquid balance sheet to finance such operations.

The Independent group was even more profitable. If its investments in the *Irish Press* and the *Sunday Tribune* had been defensive (and expensive) moves disguised, or perceived by its opponents, as ruthless expansionism, the group was simultaneously engaged in a wide-ranging series of moves outside Ireland that were in the process of turning it into a substantial international media conglomerate. Its chairman, Tony O'Reilly, whose eye for business opportunities and skill at financial engineering had made him into one of the wealthiest – and most controversial – Irish entrepreneurs (O'Toole, 1996: 63), oversaw an expansion of the company that exceeded all previous initiatives by a substantial margin.

Between 1990 and 1995 the group broadened its media portfolio in a remarkable series of manoeuvres and acquisitions. In 1991 it was unsuccessful in a bid for the Fairfax publishing empire in Australia, despite having successfully engaged the goodwill of the then Australian prime minister, Bob Hawke; this led to a long and costly lawsuit. Virtually the only other defeat the group suffered was in an attempt to launch a scratch-card system in the UK in the early 1990s, which lost an estimated £7 million. These expensive ventures apart, however, its acquisitions added considerably to its asset base, on occasion producing windfall profits, as in 1992, when its raid on *Daily Mirror* shares (it acquired £8 million worth) was seen off by UK investors and Independent Newspapers was refused a seat on the *Mirror* board: it withdrew from the battle, selling its shareholding (whose value had been pushed up by the battle for control) for a paper profit of £3 million.

Throughout 1993 the group extended its ownership of Australian local and regional titles, and formed a network of local radio stations in that country. However, 1994 was probably its most active year. In January, following its rebuff by the *Mirror,* it moved on Newspaper Publishing. This company,

which published the London *Independent* and *Independent on Sunday*, had originally been founded by a group of European broadsheets and journalists, with
the intention of providing a challenging alternative to the London *Times* and
even to the *Guardian*. The titles, however, were not doing well, and as original
investors withdrew, the Independent moved in, acquiring a 43 per cent share
for a total investment of some £9 million. Here, too, it was met with considerable hostility. The *Mirror*, which now controlled a 46 per cent shareholding,
initially refused to offer Independent Newspapers a board seat. One executive
commented: 'We won't invite anyone to dinner if they kick down the front
door.' The *Mirror* eventually ceded control to O'Reilly, although the titles
continued to lose money.

In February 1994 O'Reilly's group bought a 31 per cent stake in the South
African Argus group, later increasing that stake to 60 per cent. Almost simultaneously, it acquired for £4.8 million Capital Newspapers – a UK chain of
local newspapers published largely in the greater London area. In 1995 it
bought into the Morton newspaper group in New Zealand, and also acquired
the New Zealand Radio Network, giving it access to 41 per cent of New
Zealand's radio advertising. It also owned an outdoor advertising company in
France, an 11 per cent share in a profitable Portuguese newspaper, ten
provincial Irish titles, a controlling interest in Newspread, one of Ireland's
two newspaper and magazine distribution companies, and a 24.9 per cent
shareholding in Drogheda Web Offset Press, printers of the *Sunday Tribune*. By
1995 it was printing, globally, 12 million papers every week.

It was also engaged in defensive strategies on several fronts. Its quick
response to the dangers posed by the weakening *Sunday Tribune* and the Irish
Press group have already been noted. In response to a perceived threat from
the UK *Daily Telegraph*, O'Reilly secured acceptance of his defensive strategy
(which included a large loan) by publicly warning the workforce in January
1994: 'Employers and unions have a stark choice. They can have brutal rationalization *à la* Wapping [where ruthless management action had defeated union
power] or they can have an equitable solution to an Irish problem.'

The acceptance by the Press group's management of his offer of support,
as has been noted, postponed rather than obviated the closure of the titles.
Plainly, having found itself in a cleft stick vis-á-vis the *Sunday Tribune*, which it
was still supporting with periodic injections of loaned working capital, the
Independent management was not going to repeat the experiment. The *Press*
titles closed, in a context in which there was increasing concern about the
diminution of media diversity and the towering presence of the Independent

group in the Irish newspaper market, particularly its indigenous sector. The minister for enterprise and employment, Richard Bruton, now declined the invitation issued by the competition authority to take action against the Independent group and, instead, established the Commission on the Newspaper Industry, under the former chief justice, Thomas Finlay, to examine all the issues concerned.

The confusion and hype generated by the battle between the Irish newspapers, and between them and the UK imports, concealed the fact that newspaper sales generally were falling. In other words, the intensity of the competition between papers generally was being exacerbated by the fact that the cake was getting smaller. Some sectors, however, were more immune than others to the vagaries of the market, but for widely differing reasons. Although more than a quarter of weekly magazines were imported from the UK (and in the important sector of women's magazines more than half were imported), indigenous ventures like *Women's Way* were holding their own, supplemented by magazines targeted at younger sectors of the market such as *Image* and *U*. The magazine *IT* (inheritor of the mantle of the nineteenth-century, and originally very staid *Irish Tatler and Sketch*) was part of a group of small consumer magazines that was acquired by Robert Maxwell's interests prior to 1991 (Truetzschler, 1991a: 2). The *RTÉ Guide*, which had moved into profit after struggling to make ends meet during the 1970s, was proving a valuable addition to the national broadcasting station's cash flow. Other niche publications like the *Irish Field* and the *Irish Medical Times* were sufficiently embedded in the local topography to be able to resist outside competition (although the *Irish Medical Times* was sold to the UK Haymarket group, and remained in UK ownership for part of the 1990s before being resold into Irish hands).

The current-affairs magazine *Magill* continued to prosper, although it had a high turnover of editors under its volatile proprietor, Vincent Browne (who was also for much of this time editor of the *Sunday Tribune*). But the success of two other publications deserves to be noted, in part because they were organic developments from previous media, and in part because, as case histories, they could hardly be more different.

One of them, launched on 6 January 1982, was the *Phoenix*, a fortnightly magazine owned by John Mulcahy, whose career in journalism, as proprietor and editor, had included twelve years as editor of the now defunct *Hibernia*, and six months as editor of the *Sunday Tribune* in its first incarnation. Modelled to some extent on the British publication *Private Eye*, and resembling to some extent the former *Hibernia* in its provision of high-grade business and company news stories, the *Phoenix* carried a mix of information, scandal and gossip that saw it exceed 20,000 sales per issue by the autumn of 1992, and its regular pagination increase from twenty-four to forty-eight. Its articles were anonymous – high-profile bylines were at this time becoming the staple of Dublin journalism – and, despite an annual bill for libel settlements and costs of between £50,000 and £100,000, it has never succumbed to a really expensive defamation action.

The intense legal care devoted to the prepublication process of the *Phoenix* magazine helped to ensure that it was handled by orthodox wholesalers and appeared on news-stands. Other publications did not so appear, either by choice or because the two distribution companies regarded them as too great a legal risk. One of these, arguably one of the most effective political period-icals, was *An Phoblacht* ('*The Republic*'), the organ of the Sinn Féin party. *An Phoblacht* was unique, not least in that it survived for many years, in one form or another, while newspapers and periodicals published by other parties appeared briefly, and disappeared leaving little trace of their existence.

An Phoblacht was itself the product of a merger. Up to 1979, the party had published two journals: one in the North, called *Republican News*; the other in the Republic, called *An Phoblacht*. As the Northern conflict intensified, the circulation of *Republican News* grew steadily. One of its contributors between 1975 and 1977 (including during the period when he was in prison) was Gerry Adams. By 1978, when its circulation was approaching 30,000, it was the target of frequent harassment by the security forces, and its offices on the Falls Road in Belfast were often raided.

By the end of that year, however, an internal power struggle within the republican movement was having a visible effect on that movement's publica-tions and, on 27 January 1979, the two publications merged into a new magazine with the title *An Phoblacht/Republican News*. The change seems to have reflected a power shift away from the orthodox nationalism of Southern republicans and towards a more socially radical version of the same ideology which had for some time been emerging in its Belfast leadership (Hickie, 1994: 34). An editorial in the last pre-merger issue of the *Republican News* noted that the merger had been agreed:

> to improve on both our reporting and analysis of the news in the North and of popular economic and social struggles in the South ... [and because of] the absolute necessity of one single united paper providing

a clear line of republican leadership ... [and] the need to overcome any
partitionist thinking which results from the British enforced division of
this country and of the Irish people. (*RN*, 20 Jan. 1979)

But the new paper did more than simply provide a bulletin board for the
republican movement and its ideas. In contrast with the lacklustre and politically vacuous tone of the press of the political parties generally, it performed
a hugely important function in alerting a wide readership to issues and events
that media disinterest or (in the case of radio and television) censorship had
kept below the political horizon. It was avidly read by politicians, senior
public servants and administrators in Dublin, Belfast and London concerned
with ongoing attempts to resolve the Northern Ireland crisis. Miniaturized
versions were smuggled into Northern Ireland prisons – notably the Maze,
outside Belfast – to keep republican prisoners abreast of political developments outside, and some prisoners smuggled their contributions out for
printing.

Its role within Sinn Féin was also substantial. It not only offered its
readership a consistent and internally coherent ongoing analysis of events in
Northern Ireland, but its sales operation was closely tied in to the organizational structure of the party. The absence of distribution costs in the normal
sense of the word (it did not need wholesalers or retailers) probably made it
a profitable venture.

The broadcasting ban under the Section 31 directives, which helped to
ensure the continuing viability of *AP/RN*, continued to dog RTÉ's footsteps
until almost the end of this period. Although there were intermittent protests
by journalists and others, its existence was accepted by and large almost as a
fact of life. There was no 'peace process' as such in existence, and even the
most publicized incident evoked no more than a transitory protest. This was
in March 1988 when an RTÉ reporter, Jenny McGeever, was covering the
journey from Dublin Airport to Northern Ireland of the funeral cortege
containing the bodies of three IRA members who had been shot by SAS men
while on a mission in Gibraltar. Ms McGeever was first suspended, and then
dismissed by RTÉ after a disciplinary hearing. Her offence was that she had
interviewed a member of Sinn Féin in Dundalk in the course of the day. The
NUJ appealed her case unsuccessfully to the European Court of Human
Rights in Strasbourg. Despite its lack of success, the case uncovered RTÉ
documentation that cast the Section 31 directives and their operation in an
even more unflattering light.

Although the Section 31 directives continued to be a major irritant and an obstacle to serious journalism, RTÉ was preoccupied throughout this period with the anticipated formal end of its broadcasting monopoly (which the pirate radio stations had already ended informally), and the consequences of this for its finances and for public-service broadcasting generally. In May 1985 the minister for communications, Jim Mitchell, appointed consultants to carry out a management study of RTÉ. This study, which was completed in September, did not deal with programming issues, but indicated that RTÉ's finances were in poor shape. As RTÉ Chairman Jim Culliton pointed out, it revealed that under existing circumstances, the station could not even 'meet its statutory financial obligations' (*IT*, 20 Sept. 1985). Radio came out of the study best. It was attracting the second highest share of a national advertising pool in Europe. Overall, however, the financial situation was such that cost-control was judged essential. The report provoked major management changes, and a sense of urgency on the part of a younger team that immediately had a huge impact both on programming quality and advertising – but principally in television rather than in radio. Home-produced programming increased by 20 per cent in the autumn 1986 schedules; many Irish programmes were at or near the top of the Television Audience Measurement (TAM) viewer ratings; by late 1987, the station had doubled its profits; and by 1988, in spite of the fact that staff numbers had been cut by almost 200, RTÉ's share of multichannel viewers had increased by some 14 per cent (Mulryan, 1988: 137–8).

Radio, however, remained a problem. If RTÉ looked to the Interim Radio Commission ('CORA', from its initials in Irish), which was established by Mitchell in the same year, it was to be disappointed. The commission, which met regularly until 1987, never published a final report, but commissioned a number of consultancies which, though their findings erred on the side of caution, indicated that legal commercial radio was a viable proposition. Even if the commission had managed to produce a final report with firm recommendations, it would not have got further than the Dáil, where the stand-off between Fine Gael and Labour on this issue continued, and, if anything, became even more bad-tempered.

In the circumstances, RTÉ moved as best it could to plug the gaps in its defences, hampered not only by its acute lack of finance, especially on the capital side, but by government indecision. It had already, on 5 November 1984, launched FM3, a classical music service that shared a wavelength with Raidió na Gaeltachta, each service taking part of the day. In mid-1987, RTÉ

Radio 2 went to 24-hour broadcasting, but it seemed almost a forlorn hope: it had fewer listeners than RTÉ Radio 1, and fewer than two of the strongest pirate stations, Sunshine and Q102. Between 1980 and 1987, its audience-share had nosedived from 43 per cent to 25 per cent. Outside Dublin, some pirate stations were causing embarrassment of a different kind, this time to the government, as they deliberately targeted Northern Ireland listeners and advertisers. By 1986, according to one estimate, pirate stations in Donegal and Louth were claiming some 14 per cent of the Belfast listeners, compared with the commercial station, Downtown Radio, which was at 17 per cent, and Radio Ulster (the BBC station) at 35 per cent (Mulryan, 1988: 135).

This was the context in which, in February 1987, the Dublin government finally fell apart, a general election was held, and Fianna Fáil – which had promised commercial radio its support as long ago as 1981 – again took power. The largest opposition party, Fine Gael, was also in favour of opening up the airwaves fully to commercial competition. In November of the same year, the new minister for communications, Ray Burke, published the texts of two bills. One of them, the broadcasting and wireless telegraphy bill, was to put an end to the pirates; the other, the sound broadcasting bill, was to establish commercial stations on a legal footing for the first time. Fine Gael voted with the government on the two measures, guaranteeing them overwhelming majorities when they passed into law the following year.

The legislation created a two-tier system, with a number of orthodox commercial stations supplemented by a network of small, community stations operating largely on a voluntary basis. The latter provision was particularly welcomed by the National Association of Community Broadcasters, an organization already in existence, which had strong local (including church) support in many areas, and which published a booklet to make what it appeared were important distinctions between the stations operated by its members and the mainstream pirate stations then popular (1988). Finally, the new structure was to be under the control of a new body – the Independent Radio and Television Commission – which had the power to award licences.

Local and regional newspapers were not prevented from joining with others in applying for broadcasting licences in their area, although the legislation made it clear that this should not be facilitated if it acted to create a monopoly in any area covered by a licence, and should in any case be limited to a maximum shareholding of 25 per cent. Other interests were subject to the same limitation: national newspapers, other radio stations, non-nationals and religious denominations.

Finally, it made provision for a new commercial television service, the franchise for which was awarded to an Irish consortium, the Windmill Lane Consortium, although it was to be quite a few years, and a number of controversies, before this service materialized.

Pirate radio stations reacted in different ways to the impending legislation. Some went off air, in order to provide a sort of cordon sanitaire between their previously illegal existence and the award – or so they hoped – of an orthodox licence to broadcast. Others amended their schedules to reduce the amount of music broadcast, and to offer a limited news and current-affairs service with which to enhance their claims to legitimacy.

There were thirteen applications for the two Dublin licences and seventy-two for the provincial ones. After a series of public hearings that concentrated on the economic viability of the proposed stations (Truetzschler, 1991b: 27), the Independent Radio and Television Commission (IRTC) awarded the two licences for Dublin, and twenty-four for provincial areas. Later, a franchise for a national commercial station was awarded to a consortium called Century Radio. The first of the new stations to go on air was the Dublin-based Capital Radio (later renamed FM104), which began broadcasting on 20 July 1989 from temporary accommodation on top of the St Stephen's Green shopping centre. The first provincial station was MWR FM in Co. Mayo, which began broadcasting on 24 July.

Probably the greatest interest, however, surrounded Century, the first station to be licensed as a direct competitor for RTÉ on a national basis. Initially, there were no plans for a national competitor for RTÉ radio; this development appears to have been a personal initiative by the minister for communications, Ray Burke. The launch of the station on 4 September 1989, however, was dogged by problems. It failed to attract high-profile presenters from the national radio station, and became engaged in a lengthy wrangle with RTÉ over the cost of using the national station's transmission links, which eventually had to be resolved by arbitration. Although it attempted to supplement its income by providing a national news service for the smaller local stations, this did not succeed in offsetting costs to any significant extent, and its financial situation rapidly became extremely difficult. At this point the minister, Mr Burke, intervened dramatically, with a new package in which he proposed that RTÉ's second radio service, 2FM, should become a cultural and educational service (thus leaving the popular-music field completely to the commercial sector), that up to a quarter of the licence fee should be diverted from RTÉ to support the public-service obligations of the commercial channels, and that RTÉ's advertising should be capped, diverting revenue from that source into the commercial sector.

Burke's proposals aroused considerable controversy, and evoked a threat by the NUJ to refuse to cover a planned European summit that was due to be hosted in Dublin by the taoiseach, Mr Haughey. After a brief but intense public discussion, the government abandoned the first two proposals but, in its 1990 Broadcasting Act, implemented the third. The permissible amount of advertising on RTÉ was reduced, from 10 per cent to 7.5 per cent of total daily programming time, and the maximum permissible in any one hour was reduced from 7.5 to 5 minutes. The act also placed an absolute maximum on RTÉ advertising revenue, in that it stipulated that this should not exceed the amount received by the station from the licence fee. In effect, this was a double limitation as, given that it controlled the level of the licence fee (and was notoriously slow to increase it), the government could now also control the total revenue of the station. Burke was later to enforce a solution to an ongoing dispute between Century and RTÉ about the price to be charged by RTÉ for Century's access to the RTÉ transmission network, which it needed to ensure countrywide reception of its signal. All these matters, which were controversial enough at the time, resurfaced dramatically more than a decade later when, in the course of the Flood tribunal's investigations, an enquiry established by parliament to look into questions of political corruption, it emerged that one of the principal backers of Century, Mr Oliver Barry, had made a political donation (although not from the radio station's funds) of £35,000 in 1989 to Ray Burke, and that Century directors had intervened directly with the minister in an attempt to get RTÉ's 2FM pop channel closed down (*IT*, 20 July 2000). The second interim report of the Flood tribunal, published in 2002, explicitly concluded that the measures favouring Century (referred to above) were primarily driven by the influence exerted over Ray Burke by Mr Barry and another Century director, Mr James Stafford (Flood Tribunal, 2002: 52–3).

The Burke proposals, although evidently aimed at succouring the ailing Century Radio, were unsuccessful. The station closed down on 19 November 1991. It had achieved a 6 per cent share of the national audience.

Elsewhere, however, stations were not closing down. A national organization of commercial radio stations, the Association of Independent Radio Stations (AIRS), had been established in 1990 to act as a lobby group, and continued to do so effectively. Twenty-two of the original twenty-five stations were still broadcasting in 1992, although revenue projections were being stretched to the limit, according to Denis O'Brien, an entrepreneur who was to build his own local-radio station in Dublin, 98FM, into a major media empire. O'Brien commented in 1992 that only two of the new stations were making a profit, while the majority were still striving to break even (O'Brien, 1992: 55). 'The local stations', he added, 'are ... demonstrating that broadcasting, without state aid or influence, can be both non-elitist and of good quality'.

The advertising cap had not only failed to save Century: it had diverted advertising revenue out of the state, to Ulster Television. This was because the cap meant that RTÉ increased its price for advertising time, to make up for having to reduce the time available for advertising, to the point at which some Irish advertisers found that RTÉ was no longer cost-effective, and took their custom across the border. UTV, which was in the extraordinary situation of being the only ITV franchisee with a larger audience outside its franchise area than in it (i.e. in the Republic than in Northern Ireland), is thought to have benefited to the extent of some £3 million in the second half of 1990 as a direct result of this legislative change. RTÉ, for its part, had to engage in further staff cuts and programming economies of about 10 per cent across the board (Truetzschler, 1991b: 33). This hidden competition – hidden in that technically Ireland had no domestic commercial television, although the competitive pressures from BBC and UTV were intense – also had an effect on serious programming. A 1992 study showed that the share of its airtime that RTÉ devoted to serious programmes, at 39 per cent, was lower than in a number of other countries (Belgium, Netherlands, France and Norway, for instance) where public-broadcasting services had direct domestic commercial rivals (Kelly, 1992: 84). By 1993, in spite of the fact that there had been no increase in the licence fee for six and a half years, home-produced material had reached 45 per cent of all transmissions, and RTÉ was winning, against increasingly stiff opposition, more than 50 per cent of the audience in the 70 per cent of homes that had access to multichannel viewing (i.e. a choice between Ireland-based and non-Ireland-based channels) (Finn, 1993: 75).

The climate of cutbacks that was affecting RTÉ during this period also had an effect – and a highly deleterious one, as it was to prove – on its involvement in another radio project, Atlantic 252. This was an extraordinary development, based on an approach that had been made to RTÉ by Radio Luxembourg as long ago as 1981. Luxembourg's initiative related to permission that Ireland had received in 1975 for a long-wave frequency – a right that it had never exercised. This eventually led to the establishment of

Radio Atlantic 252, physically based in Tara, Co. Meath, but carefully concealing its actual location from the British audiences to which it was largely directed (phone-in numbers given on air were generally UK freephone numbers) and broadcasting a playlist drawn from currently best-selling music (Kenny, 1998).

The problem for RTÉ was that the complex financial arrangements under which it went into business with the other sponsor, the Luxembourg multi-national CLT (later part of the Bertelsmann-owned RTL Group), exposed it to substantial losses during the start-up period. Panicking somewhat in 1992, it sold two-thirds of its majority shareholding to help it balance its books, and was later to sell its remaining 20 per cent, again under pressure of financial stringency. By this time, however, Atlantic 252 was in the process of becoming extraordinarily successful: by 1994, it had reached a listenership of 4.9 million in Britain, higher than Radio 5 Live, as commercial radio in Britain overtook the BBC for the first time since the establishment of the commercial sector twenty-one years earlier (Foley, 1994). RTÉ was not to share in the fruits of Atlantic 252's success.

One contemporary verdict on these developments suggested that they showed that:

> Ireland has no explicit or coherent national broadcasting policy. Developments in broadcasting tend to proceed with little or no public debate and to be in line with the interests of the members of the various elites in Irish society. The wider social and cultural implications for Irish society of changes in broadcasting are rarely debated – in fact decisions are made and justified in line with economic and commercial criteria. (Truetzschler 1991b: 33)

The evidence adduced above suggests that this was a rather narrow verdict. The lineaments of a policy could in fact be discerned principally in the 1988 legislation – legislation that plainly envisaged the emergence of a strong private sector as a countervailing influence to the untrustworthy and overly critical (from the point of view of the largest political party in the state) public-service agenda of RTÉ.

That policy was itself a contested one, as was shown when a new government took office in January 1993. This government was composed, unusually, of the Labour Party, which had traditionally favoured a public-service broadcasting agenda, and Fianna Fáil, which had opposed it.

Moreover, the new administration placed responsibility for elements of broadcasting policy into the hands of Labour TD Michael D. Higgins, who oversaw the new department of arts, culture and the Gaeltacht. In a move that clearly countered the pro-commercial thrust of the Ray Burke era, Higgins immediately introduced the 1993 Broadcasting (Amendment) Act, which removed the 'cap' on RTÉ's advertising revenues. However, he used the same piece of legislation to simultaneously innovate a new policy direction: he introduced a requirement for RTÉ to set aside a figure equivalent to 20 per cent of television-programme expenditure to commission the production of content from independent production companies. Although the policy reflected Higgins' broader ambition to develop the Irish audiovisual sector beyond RTÉ (in the same year he re-established the Irish Film Board), the measure also responded to an influence that would increasingly make itself felt in Irish broadcasting in the twenty-first century: the European Union. In 1989, the European Commission had introduced the Television Without Frontiers directive, a measure designed to create a single European media market. The primary means of achieving this were a prohibition on member states blocking broadcasts from other EU territories and a requirement (albeit one undermined by the phrase 'where practicable') for EU broadcasters to reserve the majority of airtime for content produced in Europe. However, in a bid to incentivize the development of the European content-production sector, the directive also required EU broadcasters to source 10 per cent of their content from independent European producers. The parallel measure in the 1993 Irish legislation was clearly influenced by that context.

The longest-lasting element of broadcasting policy, irrespective of party, had been the various directives issued under the original Section 31 of the 1961 act. These remained in force until January 1994, when Higgins secured the government's agreement for a proposal not to renew the ban on the organizations named in the directives. From this time onwards, therefore, RTÉ was free to interview across a much wider political spectrum, as were the BBC and ITV, whose restrictions were similarly relaxed, no doubt as a result of co-ordination between Dublin and London. There were to be no dramatic developments, as the RTÉ guidelines were further revised to deal with the new situation, but an appearance on the *Late Late Show* by Gerry Adams, president of Sinn Féin, marked the watershed in a significant, and predictably controversial fashion, as he was subjected to uncharacteristic coolness from then host Gay Byrne, and to sustained criticism from four panellists. The ending of the ban had been widely touted as a confidence-building measure in support of the peace process, and was so to prove when, on 31 August 1994, the IRA announced its ceasefire.

In Northern Ireland's media developments during this period, newspaper changes were slight. In a portent of what was to come, however, a Scottish media company, Scottish Media Holdings, bought into the profitable Morton local newspaper group, establishing a bridgehead that was to have substantial consequences for the pattern of media ownership on both sides of the border. Structural changes in broadcasting were also relatively minor. Commercial radio had arrived on 17 March 1976 with Downtown Radio: it was the seventeenth local commercial station in the UK to go on air. In 1996 the station was taken over by Scottish Radio Holdings, and had a cumulative weekly reach of 587,000. There were two smaller commercial stations: City Beat, in Belfast, and Q102, based in Derry. The BBC's only Northern Irish station focused on a specific urban market, Radio Foyle, went on air in Derry on 11 September 1979.

Policy battles, particularly about editorial matters, continued to generate considerable attention and were increasingly involving the central authorities of both networks. One of the two major controversies in this period was the 1985 BBC documentary 'At the Edge of the Union', which in depicting Sinn Féin's Martin McGuinness as someone leading a fundamentally ordinary and non-threatening lifestyle provoked (despite the insertion of a thirteen-year-old film clip depicting the results of IRA violence in Belfast) a major controversy (Butler, 1995: 39). An equally critical clash came in 1988, when the ITV franchise Thames Television prepared a documentary for its *This Week* slot on the death of three IRA members in Gibraltar at the hands of British armed forces. The British foreign secretary objected strongly to the proposal to screen the documentary, which was nonetheless broadcast on 29 April, touching off a major controversy involving television executives at all levels and the British prime minister, Mrs Thatcher.

Almost immediately the British home secretary used his powers under the 1981 Broadcasting Act and in the BBC's licence and agreement of the same year to ban all direct interviews with members of Sinn Féin and a number of specified paramilitary organizations, both republican and loyalist. The situation in Britain and Northern Ireland now paralleled, almost exactly, that which had obtained in the Republic since 1972. Two years later, the 1990 Broadcasting Act replaced the Independent Broadcasting Authority, which had done its best to defend the programme, with the Independent Television Commission, and with a much more mercenary approach to the whole question of franchising (Fleck, 1995: 10). Under the new dispensation Thames lost its contract as an ITV franchise to Carlton, which distinguished itself later in the 1990s by broadcasting a number of documentaries based in part on sources whose reliability was an issue.

One of the consequences of this was what has been described as the turn towards the drama-documentary, such as Granada Television's 'Who Bombed Birmingham?' (1990). One commentator noted:

> As the ability to make factual programmes decreased journalists turned to dramatizations, which offer increased space for dramatic licence and make it easier to represent events without requiring informants to appear on television. (Miller, 1994: 275)

The same author suggests (1994: 276) that the overall performance of the media can best be described as contradictory. The media, for example, may have helped (evidently at different times and in different ways) both in creating the climate of opinion that enveloped and probably influenced the wrongful conviction of the defendants who were charged with the 1974 Guilford and Birmingham bombing, and their subsequent release after it had been shown that their convictions were unsafe and unsatisfactory. The legal process that ended their imprisonment was initiated, at least in part, as a result of a series of documentaries made by Granada and Yorkshire Television between 1984 and 1987, as well as another by Granada in 1990, which went so far as to name the actual culprits.

In the course of this process, the simplistic arguments of the early 1970s about censorship had given way to conflict and self-examination on the part of many broadcasters and administrators. But there was little in common between the often oversimplified analyses that saw each contested incident either as another example of Saxon perfidy, or as an argument in favour of the view that people who engaged in or supported armed insurrections against the crown thereby forfeited all their civil rights, especially the right to appear on national television. The complex questions about the role of the media in a civil society where the predominant mode was conflict rather that consensus were touched on only intermittently, and it is difficult in the end not to sympathize with David Butler's verdict:

> In general, coverage of Northern Ireland displays a depressing dependence on second-hand motifs, visual and thematic, which mars all

but the most painstaking and imaginative representations. And this is also the root of the problem for broadcasters, the makers of fiction and documentary films and academic analysis alike: how does one go about representing 'culture and identity in Northern Ireland' in ways which avoid depoliticizing their seamier aspects while at the same time not falling into the trap of reliance on cliché? (Butler, 1995: 44).

8 Local, national, global, 1995–2000

The last decade of the twentieth century, and in particular its last six years, saw an acceleration of the trends that had already become evident in media over the preceding decade. As the economy improved dramatically from 1995 onwards, media profits improved, media companies changed hands with increasing frequency as the tempo of rationalization increased, new titles appeared, and globalization began to make its presence felt. Broadcasting was in flux: the new commercial radio stations were taking audience-share from the only possible source available – RTÉ – and the launch of a national commercial-television competitor for the public-service station, although delayed by what appeared to be interminable wrangles and financial difficulties, finally materialized.

Little of this was at first apparent. The Irish Press group had collapsed in the summer of 1995. The role of the Independent group in its demise was the subject of much speculation, as was the now enhanced position of that group, particularly in the indigenous media sector. In the relatively few newspapers now left outside the Independent group's direct or indirect sphere of influence, concern was being expressed about the reduction in the number of media outlets resulting from the closure of the Press group. Simultaneously (and not entirely consistently), all the print media, through National Newspapers of Ireland, were putting pressure on the government to take action against the overspill of the British print media into Ireland. The paradoxes inherent in this situation were neatly exemplified in the decision of the *Cork Examiner* – a member of NNI – to take on a large and lucrative printing contract for sections of the Irish edition of the UK *Sunday Times*, which was of course in direct competition with a number of papers published by other members of the NNI.

The government's response was to set up a commission – the Commission on the Newspaper Industry – which was established in September 1995 by the minister for enterprise and employment, Richard Bruton, with a former chief justice, Mr Tom Finlay, as its chairman. If this was intended to buy time for the government, it did not buy much. The commission, despite its unwieldy size (it had twenty-one members representing almost every conceivable interest with the exception of the print unions), reported on its twelve wide-ranging terms of reference in less than a year, on 24 June 1996.

Its deliberations addressed a number of core concerns, all of them to some degree inter-related: the perceived danger of a threat to diversity of opinion caused by the demise of the Press group and the increasing role of the Independent group; the dangers posed by the overspill of UK titles; the competitiveness of the industry generally; and legal and taxation issues.

The commission's report (CNI, 1996) adopted the industry agenda almost entirely in a limited number of areas. On taxation, for instance, it supported without reservation the industry's plea for an amelioration of the rate of VAT on newspapers (although it recommended a zero rate, in ignorance of the fact that this was not permissible under EU regulations). It also adopted, and to some extent refined, the findings of the 1994 Law Reform Commission report on defamation, suggesting a number of measures that would have the dual effect of making it possible for genuinely aggrieved people to obtain low-cost remedies from the media, while at the same time protecting newspapers, radio and television to some extent against opportunist claimants. It responded in an even-handed way to allegations that the media were guilty of invasions of privacy too frequently, by accepting that while privacy was important, there were always circumstances in which the public interest demanded that it be given a lower priority.

In a number of other areas, however, the CNI fell somewhat short of the industry's expectations. While it accepted that the closure of the Press group posed distinct problems for newspapers in Ireland, it suggested that some of them were of the industry's own making. The report was, in turn, reviewed by the competition and mergers review group of the department of enterprise, trade and employment, which rejected the idea that a specific protection, by way of definition, for 'indigenous' newspapers should be inserted into competition legislation, not least because it might 'amount to a form of discrimination (even if indirect) on grounds of nationality' (Competition and Mergers Review Group, 2000: 254).

Certainly 1995 was a low point in the fortunes of Irish newspapers generally. Their share of the total market had dropped by 7 per cent in the five-year period from 1990, precisely the share by which the UK-owned titles had increased. They held 69 per cent of this market overall, and only 64 per cent of the hotly contested Sunday market. These figures, however, tended to mask another trend: the ongoing overall decline of newspaper ciculations. Between 1990 and 1995, the total number of papers sold weekly in the state had gone down from 5.6 million to 5.0 million, a substantial decrease. In this context, the main problem, as identified by the CNI, was not the threat from UK papers as such, but the threat from a generally declining newspaper readership. The commission went even further, in that it identified the high cost base of Irish newspapers as a factor that made them more vulnerable to UK competition (where massive economies of scale of course applied), and hinted strongly that remedial steps in this area were primarily the responsibility of Irish newspapers.

The commission also strongly recommended that, insofar as declining newspaper circulations could be seen to reflect public dissatisfaction with their activities and standards, newspapers could also play a part in rectifying this situation by co-operating on a voluntary basis in the establishment of an industry-wide ombudsman (the commission rejected the idea of a press council as being too unwieldy in a country of Ireland's size), who would be statutorily protected against any actions for defamation arising out of the performance of his duties. This proposal was not universally popular among newspaper-management representatives on the commission, but received substantial and, in the event, effective support from the other members.

The central dilemma faced by the commission, which it was unable satisfactorily to resolve, was that one of the problems with which it was attempting to deal – the diminution in the diversity of Irish print media, reflected in the ever-increasing power of the Independent group – could be interpreted (and, by the Independent group, was consistently interpreted) as a legitimate, indeed inevitable, response to the other problem with which the commission was grappling: the threat from imported titles.

In this context, the Independent group interpreted the Irish newspaper market as meaning the entire market, including imported titles. In this market, they argued with some justification, their presence, although substantial, was not a dominant one within the somewhat technical definition of that term in competition legislation. They also argued – and on the face of it this was not unreasonable either – that one of the best defences against foreign predators was national support for policies designed to encourage the growth and development of strong indigenous newspaper interests (such as the Independent group). They should not, they argued, be penalized for success, especially if the outcome of such a policy would be a weakening of the capacity of Irish newspapers to counter UK-based competition. And they maintained that diversity of viewpoint could be supplied as readily by diversity of titles as by diversity of ownership.

Their critics preferred to define the indigenous Irish newspaper market as the central location for the argument, and maintained that, within this more restricted arena, the presence and size of the Independent group was now becoming an effective barrier to diversity. The group, at the time the report was written, owned or had substantial interests in seven out of the eleven indigenous morning, evening and Sunday newspapers. Its size could also be interpreted as a potential source of danger. The Independent group, although a multinational with considerable assets, was not necessarily immune, by virtue of that fact alone, from predators – in fact, its success and profitability might, in certain circumstances, make even it an attractive takeover target. Nor could it be protected by national legislation from such a threat, if it ever emerged. Ireland's membership of the European Community precluded such a course of action, at least in relation to initiatives originating in another European Community country.

The commission concluded on this issue that there was in the Republic generally 'a sufficient plurality of ownership and of title to maintain an adequate diversity of editorial viewpoint and of cultural content', but it added that:

> Considering the indigenous industry separately, the recent disappearance of the Irish Press titles has involved an unwelcome reduction in diversity. Nonetheless the Commission believes that the existing industry and in particular the indigenous sector of it provides a valuable representation of contemporary Irish culture. The Commission would be concerned that any further reduction of titles or increase in concentration of ownership in the indigenous industry could severely curtail the diversity requisite to maintain a vigorous democracy. (Commission on the Newspaper Industry, 1996: 30)

The qualified gloom of the commission's report echoed its times: 1995, when it was composed, was by any standard a difficult year for Irish newspapers. Even in 1995, however, newspaper profits in all the major groups apart from the Press were healthy and improving. Subsequent events were to indicate that this particular year had, in fact, marked the bottom of the curve in what has always been a cyclical business.

The critical factor was, undoubtedly, the state of the economy rather than any measures taken specifically by, or on behalf of, newspapers themselves. The five-year period 1995–2000 saw an economic upswing in Ireland. This led to a reduction in national debt, the disappearance of emigration (and the beginnings of immigration), and associated rises in advertising revenue for all major media. Advertising in newspapers that were members of NNI, which was running at £91.9 million in 1995, almost doubled by 1999 to £181.3 million.

Sales of daily newspapers grew by 93,000 over the 1990–9 period, notwithstanding the loss of 60,000 sales by the *Irish Press* in 1995; in other words, the total market grew by 55 per cent more than the sales lost through the *Press'* demise (Barrett, 2000: 3). UK titles, it is true, increased by marginally more than indigenous titles, but the bulk of this increase was accounted for by one title – the *Sun*. In the Sunday market, Irish titles actually recovered market share in the period since 1995, with increased sales of 70,000 copies compared to a reduction of 26,000 in imports. These trends were favoured by price stability: Irish daily and Sunday newspapers, which were traditionally (especially the daily papers) more expensive than the UK imports, managed to hold their prices steady for almost all of this period, only increasing them in 2000.

Existing titles and a new title contributed in different ways to this improved market share on Sundays. The most dramatic performance, in percentage terms, was registered by the *Sunday Business Post*, which went from 33,000 in 1995 to almost 50,000 at the end of 1999. Along the way, it changed hands: in 1997 both the German investor and the original journalist-owners were bought out by Trinity Holdings, the UK newspaper group that was then engaged in a massive expansionary programme. The new title was *Ireland on Sunday*, which began life as *The Title* on 28 July 1996. *The Title* was an attempt to launch a Sunday paper that would cover only sport, made by a group of businessmen, some of whom had interests in printing. It was a brave and colourful effort, but a gamble that did not quite come off. After little more than a year, it was transformed into a new publication, *Ireland on Sunday*, which appeared for the first time on 21 September 1997. By the end of 1999 *Ireland on Sunday* was selling more than 65,000 copies, its editorial content a mixture designed to appeal at least in part to the slightly more conservative, more nationalist readers who had been left high and dry by the collapse of the *Sunday Press*.

Ireland on Sunday's finances, however, were never entirely secure, and it was taken over in June 2000 by Scottish Radio Holdings, whose presence on the Irish media landscape was becoming a significant one (Felle, 2012). It had already in 1998 purchased the regional newspaper the *Leitrim Observer*, and now added not only *Ireland on Sunday* to its portfolio, but two other regional titles, the *Midland Tribune* and the *Tullamore Tribune*, on whose presses *Ireland on Sunday* was printed. By the end of the decade, *Ireland on Sunday* had a circulation that was peaking at around 70,000. Some 17,000 of those sales were in Northern Ireland, and under the new SRH management, plans were being made to print a more fully editionized Northern Ireland edition at the Morton Newspapers plant in Portadown, which was also owned now by SRH.

On the import side, the most spectacular growth was registered by the *Sunday Times*, which opened permanent offices in Dublin in 1996, and which by the end of 1999, had increased its circulation in the Republic from 59,000 (in 1995) to 89,000.

The one black spot – and not only in Ireland – was, and remained, the evening-newspaper market, where the continuing weakness of all titles tended to undermine the other newspapers' argument that their main difficulty was posed by UK competition. Between 1990 and 1999, sales in this sector declined by some 522,000 a week (Barrett, 2000: 5). In this context, one of the bravest – or most foolhardy – decisions was that by a group of investors to launch a new evening newspaper, the *Evening News*. It was even more ill-advised to launch it in June 1996, just on the edge of the holiday season, when there is traditionally a dearth of advertising, and to have it printed in Tullamore, 64 miles away from its major potential market in Dublin. A shortage of working capital was compounded by investment decisions that favoured the purchase of expensive technology over marketing. It closed in September of the same year.

Three of the four daily papers were the primary contributors to the resilience in the morning daily market: the *Irish Independent*, the *Star* (which was part-owned by the Independent group) and the *Irish Times* all put on substantial circulation during this period. Proportionately, though, the biggest gainer at the upper end of the market was the *Irish Times*, which between 1995 and the end of 1999 added almost 17,000 copies daily, compared with the growth in the *Independent*'s circulation of some 11,600. At 113,800, the *Irish Times* was now 51,000 copies behind the *Irish Independent*, compared with some 57,000 in 1995, when the *Independent* had benefited to a greater extent from the closure of the Press group's titles.

Circulation of the newspapers generally was enhanced by the political volatility of the 1990s, and by the development of an even more adversarial relationship between journalists and governments. Three journalists secured substantial financial damages from the government because their phones had been tapped over lengthy periods going back as far as the 1970s. As already noted, two of them were Bruce Arnold of the *Irish Independent* and Geraldine Kennedy of the *Irish Times*. Formal authorization had been secured for the taps placed between 1975 and 1983 on the telephone of the third journalist, Vincent Browne, whose settlement was announced only in July 1995, but the government agreed in this case that there had been no justification for the taps and that they had been used for improper purposes. Another factor was the coming into force of the Freedom of Information Act on 21 April 1996, which gave Irish journalists a new weapon, which they used, with more enthusiasm and to greater effect than their colleagues in other jurisdictions with similar legislation, continually to embarrass both governments and public servants on issues of current controversy.

This was a period during which journalists themselves were also becoming news. The most dramatic example of this was the case of Veronica Guerin, who had begun her journalistic career on the *Sunday Business Post*, and then moved, via the *Sunday Tribune*, to the *Sunday Independent*, where she specialized in crime reporting. She was threatened on more than one occasion by members of drug gangs about whose activities she was writing regularly, and was the victim of one unsuccessful assassination attempt. Then, on 26 June 1996, aged 37, she was shot dead by a gunman on a motorcycle as she sat in her car at a set of traffic lights on the outskirts of Dublin. Her death caused profound shock, but not always for the obvious reasons. Some of her friends alleged that her employers had failed to give her sufficient protection in what was becoming known as a difficult and dangerous calling; the Independent group responded that she had refused all offers of protection, and set up an award in her honour. A Dublin criminal was later sentenced for her killing, but questions remained – about the lengths to which journalists should go in any given set of circumstances, and about the murky world in which police and criminals traded information about each other to journalists, and used publicity as a weapon.

An almost equally dramatic series of events surrounded an attempt in 1999 by the British authorities to force Ed Moloney, the Northern editor of the Dublin-based *Sunday Tribune*, to surrender to the security forces notes he had made of an interview with a murder suspect in 1989. An initial court ruling, in September 1999, went against him, but he won a landmark decision on appeal on 27 October. Mr Justice Carswell, in the court of appeal, while declining to accept that Moloney (or any journalist) had a right to withhold sources of information at will, and even though he accepted that Moloney's notes might have been useful to the police, concluded that the police had not produced persuasive evidence that they were essential, and in effect declined to authorize a fishing expedition. This judgment was warmly welcomed by the National Union of Journalists as a significant step towards greater protection for journalists' sources.

Increasing competition between newspapers led to a renewed focus on exclusive stories. These varied in quality. At the tabloid end of the market, the exclusives tended, increasingly, to be more about the private lives of public figures – politicians, musicians and television personalities – than about major social and economic issues.

This in turn prompted a renewed debate about journalistic standards, although in a different context to that in which it had been advanced at the time of the abortion controversies of the early 1980s. In the privately owned print media, the debate tended to be about the degree to which a public person's private life could become a legitimate matter of public interest. This threw up sharply differing viewpoints, as in the 1999 controversy about two cabinet ministers who had taken holidays in a French villa owned by a major Irish business figure. The minister for finance, one of those involved, bluntly told reporters that where he took his holidays was none of their business, which only made them redouble their attention.

Although it is difficult to disentangle cause and effect, the growing public (and not just journalistic) interest in media matters was further evidenced by the introduction of regular media pages in the two leading national newspapers, the *Irish Times* and the *Irish Independent*. Much of this material, particularly in the *Irish Times*, was aimed at schoolchildren. The syllabus in post-primary schools was being widened to take in the study of modern media, especially in Ireland, as a key element. Educational institutions were also adapting: by the end of the century, public undergraduate and post-graduate education in journalism was already well established (Dublin City University 1982, University College Galway 1986, Dublin Institute of Technology 1995), and was complemented by courses offered in a number of private colleges.

In the area of public-service journalism, attention turned more to RTÉ, and to critiques that it was adopting a liberal consensus to the detriment of its public-service, pluralist role. In 1999, it lost an important constitutional action taken by a private citizen who persuaded the supreme court that it had unfairly allocated more time in party political broadcasts preceding the 1996 refer-endum on divorce to the pro-divorce argument than to its opponents. RTÉ argued unavailingly that, since most of the recognized political parties were formally in favour of introducing divorce, they were simply reflecting public opinion. In June 2000, there was another controversy, this one surrounding RTÉ's coverage of the appointment, by the government, of a former supreme court judge to a position with the European Investment Bank. On that occasion, RTÉ received formal notification from the judge's legal advisers that they considered the station to be in breach of its legal requirement to be impartial on all matters of public controversy, and demanding – instead of recourse to the broadcasting complaints commission – an impartial enquiry into the whole matter. The historian J.J. Lee, in no sense an enemy of public-service broadcasting, suggested in this general context that imported television culture, replicated to some extent in the national station, risked promoting a 'consensus in favour of the principle of an atomized society based on the value of no-fault individualism' (Lee, 1997: 18).

These controversies, if anything, helped to contribute to media growth and profitability. The success of the *Irish Times* in colonizing its niche, and its apparent encroachment on the *Irish Independent*'s middle-market dominance, however, masked a problem that continued to restrict its opportunities for growth. This was the fact that although the *Irish Times* could offer advertisers not only colour but a high penetration of the upper socio-economic readership groups, the Independent group could offer huge numbers of readers across a wider social spectrum, not least by means of discount packages across some or all of the various titles it controlled. This prompted the management of the *Irish Times* and the management of the *Cork Examiner*, at various times in the late 1990s, each to consider the introduction of a Sunday paper, which would give them a greater pool of readers to offer advertisers, as well as better utilizing their relatively under-used printing capacity. Neither company took the risk, which was probably wise.

The *Independent*, meanwhile, was conscious of the competition, and moved simultaneously in two ways to keep it at a distance. The first was its decision to launch, on 15 November 1997, a colour supplement with its Saturday edition. (It was announced as a free supplement on the day of first publication, but a note elsewhere in the same issue of the paper said the cover price had been increased by 5p. to cover higher printing costs.) This addition, with part of its editorial material sourced from the *Daily Mail* and a seven-day television-programme section, certainly helped to keep the competition at bay. The second was to take action on a long-delayed decision to invest in new printing equipment. This involved a move to a purpose-built £60 million printing plant on the outskirts of Dublin, with excellent motorway access, which was commissioned in the summer of 2000, giving the *Independent* a hugely improved capacity for colour and more reliable printing generally. The *Irish Times* was also in the process of moving, also to a new, purpose-built plant in the same part of the city, but at a slower pace. Both newspapers would maintain their centre-city locations for editorial and commercial purposes.

Although the Independent group's profitability was by now legendary, it was still experiencing difficulties in another area – one judged critical to its future growth. This was its investment in new media, specifically its involvement in Independent Wireless Cable, the company that it had originally formed for the purpose of applying for a number of franchises in the television-distribution MMDS network in the late 1980s. The MMDS system was designed for areas where cable relay systems are too costly, and had substantial potential for telephony as well as television. In September 1989, the minister for communications, Ray Burke, announced the award of seven MMDS franchises to Independent Wireless and Cable, although these were largely in areas where viewers could already pick up UK signals with relative ease. The company also had minority interests in two companies in Limerick and Cork, which also received franchises.

Two years later, in a climate of some anxiety caused by the government's inability or unwillingness to deal with the problem caused by the operators of deflector systems that illegally rebroadcast cable television signals beyond the range of the official cable franchises, Burke issued a 'letter of comfort' to the Independent group, in which he assured it that once MMDS was available in any franchise region, his department would apply 'the full rigour of the law' to the illegal operators (*IT*, 30 May 1998). It was, in a sense, the pirate-radio scenario all over again. The illegal operations were popular, and therefore the government could not risk moving against them unless something realistic and acceptable was poised to take their place.

Matters reached a head at the climax of the general election campaign in June 1997, when the *Irish Independent*, on the morning before polling day, published a front-page editorial urging its readers to vote for the Fianna Fáil/Progressive Democrats opposition. It was unusual enough for this paper to publish a front-page editorial. To publish one urging such a course of action on its readers, on the part of a paper that had been sceptical of Fianna Fáil at best for most of its existence, or at least until 1979, was unprecedented. The editorial argued its case on economics and taxation policy alone. Taxpayers, it maintained, had suffered enough under the so-called 'Rainbow Coalition' of the Fine Gael, Labour and Progressive Left parties, which had held power for the previous three years. Now it was 'payback time' (*II*, 5 June 1997).

When the votes were counted, it emerged that Fianna Fáil and their allies, the Progressive Democrats, had a large enough number of seats to form the next government with the support of a number of independents. The brouhaha about the supposed role of the *Irish Independent* in assisting this result to some extent obscured the fact that the Fianna Fáil vote was the second lowest in that party's seventy-year history, and that its seat total was due primarily to highly effective vote management under the proportional representation system (which Fianna Fáil had twice unsuccessfully tried to abolish). The *Irish Independent* received a huge number of protesting telephone calls from its readership, and the controversy was reignited dramatically within days when the *Irish Times* published extracts from government documents indicating that, at a meeting with government ministers on the MMDS issue before the election, representatives of Princes Holdings – an Independent group vehicle, of which Independent Wireless Cable was a subsidiary – had indicated that failure to resolve the MMDS impasse would result in a loss of goodwill towards the government.

The implication of the conjunction of these facts was unmistakeable: that the *Irish Independent* had threatened the government that its titles would be used against them politically if the government refused to take the necessary steps to sort out the mess. This interpretation, however, was emphatically rejected by the Independent group, which maintained that there was another, more benign explanation. The meeting and the editorial, they argued, were unconnected. The threat (if this is what it was) at the meeting was made in a different context entirely – the increasing pressure from their co-investors in Princes Holdings to bring a court action against the government to establish their rights and force the closure of the illegal relay stations. Up to that point – the Independent version had it – they had been successful in persuading their foreign investors that the problem could be resolved by negotiations. If negotiations proved fruitless, they would no longer be able to argue convincingly against legal action and the government would, for this reason, have forfeited their goodwill.

There was a flurry of statements and counter-statements. The editor of the *Irish Independent*, Vincent Doyle, threatened to sue the *Irish Times* for defamation, and the sudden drop in temperature in the relationship between the two newspapers, situated on opposite sides of the river Liffey, was palpable. It blew over but, whatever its origins and intentions, the editorial marked a watershed in Irish journalism that few would forget.

It was in the same year, 1997, that the *Independent* first went online. It was not the first Irish newspaper to establish a website – that honour went to the *Irish Emigrant,* which was established in 1987 as an entirely net-based newspaper, and which was so successful that it found itself producing a hard-copy edition for US readers in 1995. In 1999, it was presented with the Net Visionary Award by the Irish Internet Association. By 2000, it had paid subscribers in 112 countries. The *Irish Times* launched its website in 1994, purchased a portal site in 1998, Ireland.com, and was consistently registering usage figures that put it among the leading newspapers in the world. Even the *Clare Champion* preceded the *Irish Independent* into cyberspace, by launching a website in 1994.

It became clear early 2000, however, that the *Independent*'s website was not the only, and probably not the primary, focus of the Independent group's venture into new media. In January, the company – renamed Independent News and Media since the previous year – bought Internet Ireland, which had been formed in 1997 to provide internet access to a large number of corporate clients. Three weeks later, on 5 February, the company announced the creation of a new venture called Unison. This initiative associated Independent News and Media directly not only with Internet Ireland, which it now owned, but with thirty-eight regional newspapers. The newspapers would promote Unison – a set-top internet connector designed to help increase Irish adult internet usage from 17 per cent to 34 per cent, or 800,000 adults, within a year. And Unison would promote the newspapers and the services they offered, in a form of commercial synergy that would be free – at least to the newspapers concerned. Each regional title participating would have one hundred free set-top boxes to give away; thereafter, they would cost some £300 apiece. There were other linkages, too: International Network News, a Dublin-based radio news service, was not only providing timed news bulletins for the regional radio stations, but was acting as part of the platform for Unison and, additionally, was involved in creating a link for news from the Republic for UTV's website in Northern Ireland.

Its move into new media did not mean that Independent News and Media had taken its eye off terrestrial opportunities. One such had been in the process of emerging since 1998, when the UK-based Mirror group had bought the weakened *News Letter* in Belfast. This was a strategic purchase, not least because it gave the Mirror group access to printing facilities in Ireland for its daily and Sunday titles. In 1999, however, the pattern of media ownership in Northern Ireland was again dramatically disturbed when Trinity Holdings, now the owners of the *Belfast Telegraph,* and its associated titles (including *Sunday Life*), made a bid for the Mirror group. This bid succeeded, creating a new media giant to be known as Trinity Mirror. But the British government demanded that Trinity sell off the *Belfast Telegraph* if it wanted to finalize the Mirror bid; had it not done so, the new company would have controlled an unacceptably high 67 per cent of the Northern Ireland print-media advertising market. This was a significant concession: the previous year, the *Telegraph* had contributed almost a quarter of Trinity Holdings' profits, making some £21 million on sales of £54 million. The new Trinity Mirror conglomerate, as it happened, retained ownership of two other Irish newspapers, titles in the Derry Journal group, which it had acquired in 1998 for £18.25 million.

As suitors circled the *Telegraph*, the Independent group said little, but it was hardly entirely unexpected when, on 29 April 2000, details of its £350 million bid for the paper were finally revealed. It had already, in anticipation of unionist objections (which were duly forthcoming, not least from the deputy leader of the Unionist Party, Mr John Taylor, himself the proprietor of a number of Northern Ireland regional papers), indicated that the *Telegraph* would be guaranteed continuation of its editorial independence. Independent News and Media, keen to defend itself against the possibility that it might be accused of exploiting a dominant position in the marketplace, gave undertakings in advance that it would not change employment structures in its Northern titles, would not change the *Telegraph*'s editorial policy, and would not introduce cross-border deals with the group's titles in the Republic (*SBP*, 18 June 2000). These were apparently sufficient to allow the sale to proceed.

The censorship laws (not referred to at all by the Commission on the Newspaper Industry) continued to make themselves felt, infrequently but controversially. In 1999, the popular listings magazine *In Dublin* was banned by the censorship board after the board had taken note of its practice of including advertisements for establishments that were in all probability brothels, although they were described as 'massage services'. The magazine succeeded in having the ban temporarily lifted by a higher court on the grounds that it had not been offered an opportunity to reply to the charges, and simultaneously sanitized its advertising pages. Its publisher was subsequently charged with offences under public order legislation relating to the advertising of brothels and prostitution (*IT*, 19 May 2000). The British men's magazine *Loaded* was removed from shops by its distributors shortly afterwards (*IT*, 22 May 2000).

INM's acquisition of the *Belfast Telegraph* confirmed a phenomenon that had already become noticeable: moves towards rationalization and acquisition in the Irish regional newspaper market generally. It was a landscape that had, until the 1990s, remained unchanged for the best part of a century. Originally a Protestant and settler phenomenon, the local or regional newspaper had, by the second half of the nineteenth century, become an increasingly Catholic and nationalist one. Catholic emancipation in 1829 and the failed rebellion of 1848, combined with increasing popular literacy and the emergence of a new Catholic middle class, were all factors that accelerated these developments. In 1855 there were 100 of them; by the beginning of 1859 there were 130; by the end of 1859 there were 140. By 1969, however, when a government committee on industrial reorganization reported on the prospects of the 41 remaining titles in an era of free trade, it was anticipated that many of them were doomed to closure.

Although these predictions were to a large extent unfulfilled, the general situation was stagnant well into the 1980s, when a substantial number of regional papers were owned by private limited companies or family partner-ships – some forty different companies, controlling the 87 per cent of regional newspaper circulation (the remainder was controlled by companies within the Independent group) (Horgan, 1986: 11).

By 1993, however, their number had risen to 53 and, despite gloomy predic-tions that they would suffer inordinately from the introduction of local commercial radio, in which some regional newspapers had a stake (*IT*, 27 Aug. 1983), they not only survived, but prospered. As this happened, they became increasingly attractive prospects for domestic and foreign predators. The *Cork Examiner*, which had already acquired the *Waterford News and Star* in 1959, moved completely outside its natural hinterland in 1995 with the purchase of the *Western People* in Co. Mayo; in 1996 it extended its ownership in Connaught by acquiring the *Sligo Weekender*, a free sheet, and in a bid to increase its circulation and its attraction to advertisers, dropped the word 'Cork' from its title. In 1999 it acquired a weekly paper in Killarney, the *Kingdom*, and was pushing its own daily paper outside Munster, for example by including a weekly Kilkenny supplement in its paper on one morning each week from mid-2000. That year it was also to relaunch itself yet again, as the *Irish Examiner*. Although each move was accompanied by an increase in circulation, the modest scale of such increases indicated the difficulties of taking on the established national media on their own doorstep, particularly in Dublin.

Rationalization was continuing apace in other areas too. The *Leinster Leader* and *Leinster Express* merged in December 1998, and, in early 2000, subsumed the *Dundalk Democrat,* a small and old-fashioned but nonetheless profitable paper (it was, at this stage, the only regional paper still devoting its front page entirely to advertisements). The *Kilkenny People,* owner of one of the most advanced printing presses in the country, was now the owner of the *Tipperary Star* and the Clonmel *Nationalist.* At the end of June, it was itself taken over by the expanding Scottish Radio Holdings group for a consideration of almost £30 million. This latest acquisition made SRH, whose cost-cutting, bottom-line-oriented management style was already becoming a byword in the other Irish enterprises it had acquired, the third largest regional newspaper group in the country. And these were only the most noticeable developments in an industry that was expanding. Its association, the Provincial Newspapers Association of Ireland, which had existed for more than twenty years as an organization mainly dedicated to handling wage negotiations with the NUJ, had become, by 1999, the National Association of Regional Newspapers (NARA), dedicated to selling advertising space in all its associated titles on a co-operative basis, and, incidentally, providing part of the platform for the Independent's Unison initiative.

The growth of the regional newspaper sector during this last decade of the century was all the more remarkable in that it was accompanied by an equally noticeable growth of free sheets, or newspapers funded entirely by advertising. By 1998 it was estimated that whereas NARN's titles sold 600,000 copies a week and were read by 2 million people, advertising in free sheets, which accounted for some £5 million of the £250 million annual advertising spend in the regional press, was still only half the UK per capita expenditure, and could be expected to increase substantially (*ST,* 18 Oct. 1998).

Northern Ireland's forty regional newspapers were more protected against rationalization and closure than their counterparts in the Republic, although not necessarily against acquisition. The purchase of Derry papers by the Mirror group has already been noted; and Mr John Taylor, the prominent unionist politician subsequently ennobled as Lord Kilclooney, would over the course of his business career assemble a portfolio of six Northern Ireland regional newspapers. The relatively higher number of regional titles in Northern Ireland, on a per capita basis, can be explained in part by the divided nature of Northern Ireland society. In the city of Derry, for instance, two different papers exist side by side, one catering almost exclusively for nationalist readers, and the other for unionist readers.

In the urban areas like Derry and Belfast, where social and political tensions were higher and housing was in any event more dramatically segregated, newspaper readership was more predictable in its characteristics and occasionally spawned strong local initiatives. An interesting example of this was the weekly *Andersonstown News*, which appeared in the west Belfast suburb of that name on 22 November 1972. The context was highly political, in that the conflict between the IRA and the security forces was particularly virulent at this point, and the *Irish News* was increasingly seen, by more militant republicans, as overly cautious and even, on occasion, unsympathetic to their nationalist agenda. The steady growth of the *Andersonstown News*, with its strongly republican agenda, continued throughout the period from then until the end of the century.

In the course of its growth, the *Andersonstown News* made one, unusual, acquisition. This was another Belfast paper, *Lá* ('*Day*'). Even in the special circumstances of Northern Ireland, the appearance, and continuing existence, of *Lá* would have been a matter for comment. It appeared for the first time on 13 August 1984, just after the death of a civilian in civil disturbances in Andersonstown, and was a cyclostyled daily news sheet written entirely in Irish and produced from an abandoned flax mill. It might be thought that its choice of language indicated an allegiance to the more extreme elements of republican ideology, but in fact it maintained, from the outset, that it was not associated with any political movement and, indeed, made consistent and occasionally successful efforts to persuade some unionists that the Irish language and culture was also a part of their heritage, and one which they could share without betraying their political principles. 'The Irish language movement must be above politics – and so must *Lá*', declared one of its founders, Eoghan Ó Néill (*Guardian*, 3 Apr. 1989). Despite a disastrous fire that destroyed most of its printing equipment in 1985, it staggered along on a shoestring (and with occasional support from the Northern Ireland office). In 1991 it became a weekly, giving Northern Ireland yet another distinction: it was the only part of the United Kingdom – perhaps the only part of the world – in which there was a weekly publication called '*Day*', and a monthly one called *Fortnight*. It was absorbed by the *Andersonstown News* organization in April 1999 and, despite fears that its agenda might become more political and less cultural under its new ownership, initially sold a modest but satisfactory 2,500 copies per issue. However, at the end of 2008, after two and a half decades, the cross-border Irish-language funding agency Foras na Gaeilge withdrew its funding of €400,000 per annum on the grounds that it could not be justified on the basis of *Lá*'s sales and readership figures. *Lá* had been proportionately more successful than *Foinse*, the Irish-language weekly that replaced the subsidized Irish-language weekly *Anois* in Dublin in October 1996, after the department of the Gaeltacht, which provides substantial financial backing for publications in Irish, had declined to renew its contract with the *Anois* publishers. *Foínse*, although it marked a distinct advance on its colourless and sloppily produced predecessor (it seems to have been the first Irish-language publication to include, in its first issue, a wine-appreciation column), struggled against public indifference and, perhaps more worryingly, the continual difficulty of finding reporters and sub-editors with sufficient command of the language to enable the production of a fully professional newspaper. In 2009, a year after *Lá's* demise, *Foínse* closed after its proprietor and Foras na Gaeilge were unable to agree on a new contract. Although it re-emerged for a period as a supplement distributed with the *Irish Independent* every week, in 2013 it ceased operating as a print title. An online-only version limped along for a further two years, shutting down in 2015.

The simultaneous ending of the broadcasting bans on paramilitary inter-viewees in the Republic and in Northern Ireland in 1994 was absorbed into both BBC and UTV practice with very little fuss. The major change, as far as the viewers were concerned, was the ending of a farcical practice introduced in October 1988 after the then British home secretary, Douglas Hird, issued an order to BBC and ITV prohibiting the broadcast of direct statements from representatives of paramilitary and proscribed political organisations in Northern Ireland. Faced with this, both broadcasters opted to dub the voices of figures like Gerry Adams with those of actors who delivered their perform-ances in sync with footage of the interviews. Not only did this look and sound absurd, but, as statements from banned organisations still reached the airwaves, the prohibition completely failed to achieve its objective. Some of the actors had perfected the accents of their alter egos to such a degree that the only noticeable difference, on television, was the disappearance of the caption that had previously informed viewers that the voice they were hearing was not actually that of the person who appeared to be speaking.

Both BBC NI and UTV had, by the end of 1999, evolved into steady state broadcasters, each of them operating under virtually the same conditions as the BBC and ITV franchisees operated in other UK regions, with the political controversies that had marked their evolution rapidly fading into the past. UTV's major characteristic remained its profitability. By 1998, its pre-tax profits were up to £12.5 million, a 50.8 per cent increase in the previous year. In the same year, it served notice that it might take legal action against RTÉ if it attempted to boost its signal into Northern Ireland. This threat was

plainly related to UTV's decision, in 1999, to mounted an aggressive advertising drive in the Republic. It argued that RTÉ, which had Irish rights for
popular soap operas like *Coronation Street*, would be in breach of copyright if
it broadcast across the border this and other programmes for which UTV
had rights in Northern Ireland. Cabletel, a US-owned group that was cabling
many Northern Ireland homes, decided in 1999 not to include RTÉ on its
cable until these legal issues had been resolved. RTÉ, of course, could make
a similar claim about the UTV overspill signal.

In the meantime, UTV was making its own forays south of the border. Its
combination of successful locally produced chat shows with popular
presenters, and access to sports and soap operas through the ITV connection,
continued to guarantee it massive audiences on both sides of the border.
Underlining its evident belief that globalization begins at home, it noted, in
its annual report for 1999, that UTV remained 'committed to our objective
of developing our service in the Republic of Ireland within the concept of
extending the accessibility of all the indigenous channels in Ireland
throughout the whole island' (UTV, 1999: 4).

BBC NI, on the other hand, was in a slightly more problematic position as
the region's only public-service broadcaster, in that the political developments, including the successful re-establishment of the power-sharing
executive in June 2000, inevitably created greater expectations on the part of
nationalists that their agendas would be given more prominence. This created
greater pressure on cultural institutions like BBC NI which, precisely because
they were not susceptible to direct political influence, were slower to adapt to
altered political circumstances.

The BBC was reluctant, for example (for reasons which do not need to be
guessed at), to take initiatives in relation to Irish-language broadcasting. To
do so risked irking both communities: unionists, because they felt that it was
inherently non-British or in some sense opening yet another unwelcome door
to the Republic and its machinations; nationalists, because they could be
predicted to believe that nothing the BBC does in this area will ever be
enough. By 2002, BBC NI was producing four hours of radio in Irish per
week, and approximately ten hours of television in Irish per year. The
problem also extended into sport. BBC NI had the Northern Ireland broadcasting rights from the Ulster Branch of the GAA, but rarely availed of these
for live transmission, preferring to broadcast edited highlights in the late
evening. There were two or three related problems here: the anticipated
reaction of unionists to airtime being squandered, as they would see it, on

such nationalist pursuits was intensified by the GAA's resolute refusal until 2001 to open its ranks to members of the security forces – specifically the RUC – on political grounds; and, finally, there was the general religious objection among sections of the unionist community, on sabbatarian grounds, to the playing or broadcast of any sport on Sundays. Rugby football, then a minority – though growing – sport within the nationalist community, but of considerable interest to unionists, received more live coverage. The implications for BBC NI and UTV in this area were implied rather than spelt out in the Belfast Agreement, which, as one nationalist commentator pointed out, specifically in relation to broadcasting, 'does not just seek a new method of political partnership in the North but sets out the template for a whole new society' (McGurk, 2000; *SBP,* 11 June 2000).

BBC NI is, however, not alone in experiencing problems in connection with Irish-language broadcasting. It has been a contested area also in the Republic for many years, as the controversies preceding and surrounding the establishment of Raidió na Gaeltachta demonstrate. The same has been true, at a possibly even more intense level (because of the greater costs involved), of the establishment of the Republic's Irish-language television service, which went on the air for the first time on 31 October 1996.

The history of this initiative, going back over thirty years, is a story partly of political inertia, partly of an ongoing argument about whether broadcasting in Irish, whether on radio or on television, should be integrated into the national public broadcasting service as seamlessly as possible, or should be given an independent existence and status (with the consequent risk, as many Irish-language enthusiasts saw it, of ghettoization). The 1965 report of the Commission on the Restoration of the Irish Language saw television, which had been established in 1961, as a prime motivator of, and location for, the language-revival movement; indeed, it had been so exercised by this prospect that it had issued an interim report in 1959, when the establishment of RTÉ was under consideration, focusing in large part on the opportunities presented by the new medium. The early years of RTÉ saw the emergence of a struggle between those who saw this as one of the station's prime aims (including a number of those who served on the RTÉ Authority in its early years) and broadcasting executives who were anxious that too great an emphasis on Irish would handicap its attempt to secure a national audience. Tensions ran high on this issue, at one stage provoking the resignation of the first chairman of the authority, the broadcaster and businessman Eamonn Andrews (Horgan, 1997: 320).

Even among those who favoured the use of the new medium for the revival, there were strong differences of opinion: radicals suggested a low-cost, almost prefabricated individual station (Doolan, Dowling, and Quinn, 1969: 221); the official language movement preferred to make inroads, insofar as this was possible, into the programming structure of the existing national station. RTÉ, for its part, generally opposed the creation of a separate channel, which would undermine its monopoly status, as it made clear in its submission to a 1977 government advisory committee on Irish-language broadcasting, whose report was never published but which recommended an internal RTÉ solution. Attempts to set up a pirate television initiative in Irish in 1980 failed, and in 1983 Bord na Gaeilge, the official government body with responsibility for promoting the language revival, voiced the opinion that a service separate from the national broadcaster would be a good idea, tacitly accepting for the first time that the internal RTÉ option was perhaps not necessarily the best one, and accepting the implicit risk of marginalization.

The pirate project was not dead, however: it surfaced briefly between 3–5 November 1987 in the Connemara Gaeltacht, when some eighteen hours of live television and video recordings were broadcast, mostly music programming and some debate on issues connected with the Irish language. Bob Quinn, a director, who was one of the authors of *Sit Down and Be Counted* (and to become a member of the RTÉ Authority in the late 1990s), was among those involved in the experiment, which was repeated, equally briefly, in December 1988. These and other initiatives prodded the government into action. A working group on Irish-language broadcasting in 1987 made suggestions similar to those that had been made ten years earlier, but no action was taken. A feasibility study was carried out by consultants in 1988 and published in 1989: it recommended that a new service should cater both for Gaeltacht and non-Gaeltacht audiences, with a special emphasis on children. Again, nothing was done, although it prompted the emergence in 1989 of a new organization, Feachtas Naisiúnta Teilifíse ('National Television Campaign'), which broadened the demands to include a television service in Irish for the whole country, and not just for the Gaeltacht areas. RTÉ output in Irish at this time was some 2 per cent of an increasing amount of airtime. In 1991 the taoiseach, Charles Haughey (who as minister for the Gaeltacht had promised in 1987 that £500,000 would be spent on a new service, but had not delivered on this commitment), told the annual meeting of his party that an Irish television service would be established in 1992, but this did not happen. In 1992, however, Maire Geoghegan-Quinn, the minister for the Gaeltacht, appointed a special adviser to take the project further, and she and her successor, Michael D. Higgins (both represented the Galway West constituency, which has a sizeable Gaeltacht population), although members of different governments, were actively involved in bringing it to completion.

Telefís na Gaeilge began broadcasting on 31 October 1996. Its viewing figures were handicapped by the fact that initially it was not available nationally, and the annual cost of some £20 million was the subject of inter-mittent but intense controversy, as was the fact that it did not – because it simply could not – broadcast entirely in the Irish language. At different times, it found creative uses for its spectrum allocation, ranging from live broadcasts of the proceedings of controversial parliamentary committees (notably in 1999, when one of these committees was focusing on malpractices in the banking system), to European football matches, to old recordings of GAA matches (on which the original commentary was in Irish in any case), and late-night classic films. Its most successful programmes included a soap opera in Irish, *Ros na Rún*, and (quite unexpectedly) a long-range weather forecast service, provided by a UK company, which attracted some 100,000 viewers nightly. By March 2000, it was attracting a daily viewership of 500,000, which, though lower than its target and still a source of criticism for those who believed that the new service was a waste of money, was respectable in the circumstances. It was broadcasting an average of nine hours of programming daily, five of them in Irish. It was produced by a staff of fifty people, with an average age of 25, the majority of them women, from its studios west of Galway. In the long term, its difficulties appeared to mirror those of the radio station, in that it had the problems of balancing the needs of its local, Irish-speaking audience and the national audience, and of appealing to young people in particular (Hourigan, 1996: 6).

The question of funding remained a difficult one. A 1991 report by an ad-hoc group commissioned by the government had recommended that the cost could be met by increasing RTÉ's permitted advertising time by half a minute in every broadcast hour. This technically simple method proposed was, however, fraught with political dangers. Plans for TV3, the new national television competitor for RTÉ, were already in hand; its first projected launch date of 1990 had fallen through, and it was having difficulties with its financial planning; any move to increase RTÉ's access to advertising revenue would imperil it further.

At the same time, RTÉ's finances were deteriorating steadily. Although the cap on the amount it was allowed to earn from advertising had been removed by the government in the Broadcasting Authority (Amendment) Act 1993, and operating profits for 1994 were almost £10 million, this concealed a loss of some £2 million on broadcasting. The overall profit was accounted for largely by ancillary revenues, such as income from the *RTÉ Guide*, money from the part-sale of Cablelink, and investments in Radio Tara, RTÉ's vehicle for its involvement in Radio Atlantic 252, and the Riverdance company, set up to market the hugely successful dance company, in which it had a 25 per cent stake. By the following year, the loss on broadcasting had risen to £7 million, and the situation was looking even more bleak.

The government's priorities at this stage in relation to broadcasting, however, were not financial. What it had in mind was a major restructuring of the electronic media, and its ideas on this subject were contained in a green paper published in 1995.

With its stress on the cultural process and the broadcasting context, the green paper focused on the 'centrifugal and centripetal pressures' affecting the development of broadcasting in an era of globalization (Department of Arts, Culture and the Gaeltacht, 1995: 131). 'What', it asked, 'is the optimum balance of power we should strive for in the multiple relationships between technologies, regulators, providers and users?' (143). Building its response to this question on foundations securely anchored in public-service broadcasting philosophy and traditions, the green paper suggested that there might be a case for 'merging the policy and regulatory functions of the RTÉ Authority and the IRTC to form one over-arching Authority' (166). This proposal, which was to be found in embryonic form in the report of the Broadcasting Review Committee twenty years earlier, was to be the subject of intense discussion and disagreement for the next seven years. It strongly supported the case for a separate television channel for Irish-language programmes, although it rejected the idea that this should be under a separate statutory authority (207).

All in all, although the document raised a large number of points for discussion, and gave an indication of government thinking on some of these issues, it was insufficiently focused as a framework for legislation and, to some extent, did not have the opportunity to consider the full impact of the new technologies, whose downstream implications were at this stage seen only indistinctly. By 1996, this framework was already being overtaken by the debate about transmission systems and their control, and new legislation was being prepared that in effect accepted that some of the green paper's ideas were now redundant. Current events were also adding to the pressure for decisions. UTV, which had been part of the refinanced TV3 consortium, withdrew from it in 1996 because it planned to concentrate instead on developing its own digital service, which could be made available to many viewers in the Republic without cable or satellite links.

The details of the legislation, which was predicated on a decision in favour of digital terrestrial television (the other options being terrestrial cable and satellite) were finally announced by the minister, Michael D. Higgins, on 18 March 1997. The key elements of his proposal were:

- a statutory definition of public-service broadcasting;
- the establishment of a broadcasting commission;
- the creation of a separate structure for Telefís na Gaeltachta;
- the control of rights to broadcast certain major sporting events; and
- the possibility of allocating start-up grants to news services.

The overriding function of the new commission, the minister announced, would be to endeavour to ensure that the number, categories, structures of, and arrangements for broadcasting best served the needs of the people of Ireland, bearing in mind their traditions, language and cultures and respecting the distinction between the roles and objectives of public-service and independent broadcasters (*IT*, 19 Mar. 1997). Public-service broadcasting was, and remained, a battlefield.

While the commercial stations argued that equity demanded they be given a share of the licence fee for their news and current affairs, the argument was being moved onto the European plane. In the teeth of powerful opposition, not least from European Commission officials whose views about the value of competition left little room for any Habermasian concept of the public space, Higgins collaborated with public broadcasters across Europe (most notably in France) in successfully pushing for the inclusion of the following protocol defining and defending public-service broadcasting into the 1995 Amsterdam Treaty:

> The provisions of the Treaty establishing the European Community shall be without prejudice to the competence of the Member States to provide for the funding of public service broadcasting insofar as such funding is granted to broadcasting organisations for the fulfilment of the public service remit as conferred, defined and organised by each member State, and insofar as such funding does not affect trading

conditions and competition in the Community to an extent which would be contrary to the common interest, while the realisation of the remit of that public service shall be taken into account.

The definition, focusing as it did on organizations and not, as the private sector would have preferred, on the public-service content of broadcasts from all stations, was vital to the interests of RTÉ and indeed of other national broadcasters within the European Community: it was a definition of public-service broadcasting as the activity of a public-service broadcaster – not as a generic activity that was common to publicly and privately owned broadcasters and that should get a subsidy from public funds regardless of its point of origin.

This definition, now given statutory expression at European level, was a valuable bulwark for public-service broadcasters. It came up for discussion again late in 1998, when the European Commission's competition directorate circulated a discussion document exploring the criteria on which public-service broadcasting might be defined to make it possible to give meaningful effect to the Amsterdam protocol. There was a brief debate on the document at European level, which ended when the document was rejected by the EU member states. Few people were under any illusions, however, that the argument was over.

The Amsterdam initiative in favour of public-service broadcasting, significant though it may have been, did not solve the financial problems of public-service broadcasting stations anywhere, and certainly not in Ireland. These were now, at least insofar as RTÉ was concerned, reaching crisis point with the inauguration of Telefís na Gaeltachta, and the requirement that this new service be subsidized by RTÉ, to an extent estimated by RTÉ of some £9 million annually. This was almost certainly a deliberate over-estimate, but money still had to be found somewhere.

Given that the option canvassed earlier – an increase in the proportion of broadcast time available to RTÉ for advertising – was politically unacceptable, there was really only one, almost equally politically unpopular, option left. This was to raise the licence fee, which was increased from £62, at which figure it had remained since March 1986, to £70 in September 1996. The fact that it took a decade for the ceaseless representations from the RTÉ authority to the government to be heard spoke for itself and – although the possibility of any connection between this and the voracious appetite of the infant Teilifís na Gaeilge (TnaG) for funds was not so much as hinted at –

the fact that the £5 million raised by the increased licence fee was almost exactly equal to the (revised) cost of the subsidy RTÉ was now required to give to the Irish-language station was, at the very least, an extraordinary coincidence.

This had solved the TnaG problem, but it did not solve RTÉ's problems. These were partly related to the structural model proposed by the government. If decisions about the future of broadcasting were to be made by a new super-commission on which both public-service broadcasters and private interests would be represented, the fear in RTÉ was that the public-service side of the table would be consistently out-voted. This would, they anticipated, lead to a further financial squeeze, erosion of the standards of public-service broadcasting, and diversion of the broadcasting audience towards the private channels. It meant, according to RTÉ's director general, Bob Collins, that 'more decisions about what Irish people watched would be made outside Ireland' (*IT*, 19 Mar. 1997).

A change of government in June 1997 provided some breathing space, and the new minister, Síle de Valera (a granddaughter of the founder of the *Irish Press*) addressed the problem of funding broadcasting again. As she was considering her options, the RTÉ Authority unveiled its own proposals, which envisaged groups of channels offering viewers tiered subscriptions. The first tier would be RTÉ1, Network 2 and TnaG (relaunched and re-baptized as TG4 in 1999). Other channels would contain the current crop of UK services, a 24-hour news channel, a catch-up channel with access to recent programmes and a pay-per-view channel. Within two months, the government decision was published as 'Irish Television Transmission in the New Millennium'. This envisaged a network of up to thirty channels of digital terrestrial television, grouped in multiplexes that would have five channels each. RTÉ would be allocated one of these, giving it three additional channels (*IT*, 24 July 1998). The bullish atmosphere engendered by the new plan, in which the government effectively took on board many of RTÉ's proposals, was underlined when the station launched yet another new service on 1 May 1999. This was Lyric FM, a serious music station, successor to a similar service known as FM3 which had shared the Raidió na Gaeltachta wavelength for part of the day before Raidió na Gaeltachta expanded its broadcasting hours.

This series of government decisions also included a solution to a problem that had dogged the RTÉ Authority since 1993. In that year, a Broadcasting Act had stipulated that RTÉ should set aside a certain proportion of its programming budget for productions from independent companies. Implementation of this proposal was controversial, not least because there was no agreement on which elements of RTÉ's programming budget were to be included in the total sum subject to the percentage, and which overheads could legitimately be excluded. The minister now proposed, instead of a percentage, a flat sum of £16 million, subject to some indexation, which achieved a measure of stability in the equation satisfactory both to RTÉ and the independent producers.

There was one critical difference from the 1995 green paper plan, in that the RTÉ Authority would continue to exist as a public-service enclave, ring-fenced to some degree from the new super-commission. Structurally, this was an advance, from RTÉ's point of view, but the question of finance was still up in the air, all the more so as de Valera had announced in June 1998 that she was not prepared to index the licence fee by linking it to inflation (*IT*, 19 June 1998). RTÉ, in these circumstances, was plainly in need of an ongoing source of finance, and the most obvious source of that finance was its physical trans-mission network. Under de Valera's proposals, this was to become the property of a new company called Digico, in which RTÉ would participate with partners from the private sector, who would be able to contribute a large part of the capital investment needed. When she introduced her broadcasting bill in 1999, she indicated that RTÉ would be given a 40 per cent share in Digico. The final resolution of the problem in June saw a decision to award RTÉ a 28 per cent stake in the company owning the transmission network, but – in a move designed to reassure other interests – RTÉ would have no share in a separate company that would be established to sell digital channels to the consumer (*IT*, 28 June 2000).

RTÉ's financial projections now began to look distinctly cloudy. It needed a strategic partner on the physical side, to help carry the cost of capital investment into the new millennium. Here, however, it was locked into an increasingly bitter conflict with the major media companies, notably multi-national ones, anxious to ensure a rapid increase in the number of outlets for the televisual products in which they had invested. It was also operating in a climate in which there was a poor prognosis for the agenda of public-service broadcasting generally, and in which much of that agenda was being dismissed as anti-competitive, special interest whingeing.

RTÉ also needed a strategic partner on the programming side. It could offer such a partner not only access to a benevolent tax regime for programming generated in Ireland, but also access, through an Irish cable channel, to major potential audiences with strong Irish cultural roots in the United States and elsewhere. It had already substantially improved its transmission facilities in areas covering Northern Ireland, where an estimated 70 per cent of viewers could receive the RTÉ television signal. For the time being, however, this prospect had to take second place to the more urgent structural and financial problems, exemplified by decisions in September 1998 to transfer responsibility for afternoon and breakfast television to private contractors. GMITV, the company to which breakfast television was subcontracted, had Bob Geldof of Planet TV as one of its investors; the company was allowed to sell the advertising time within its programme slot. Although, in the event, the GMITV deal did not proceed, the putative deal illustrated the urgency of RTÉ's need to raise funds.

The financial problems were being exacerbated by the relaunch of national commercial radio and the successful launch of TV3. National commercial radio had been in the doldrums since the collapse of Century in 1991. By late 1995, however, the IRTC had invited new tenders for a replacement station and, in February 1996, the franchise was awarded to a new consortium that included the progenitors of the Riverdance company, the *Irish Farmers Journal* and the *Cork Examiner.* One of the unsuccessful applicants was a consortium that include the *Irish Times* and Richard Branson's Virgin company (*IT,* 9 Feb. 1996).

The launch of the new commercial radio station on 17 March 1997 was mired in controversy. Its chief executive, Dan Collins, who had been recruited from one of the more successful regional stations, Radio Kerry, was dismissed. The *Irish Farmers Journal* withdrew its investment because of fears that its status as a charity might be compromised (the *Cork Examiner* had withdrawn in February, citing policy differences with the IRTC). In June, three months after it had gone on air, three-quarters of the national radio audience were saying that they had never heard of it, and there was an ongoing dispute with RTÉ about the cost of the physical links on which the service was being distributed.

The withdrawal of the *Irish Farmers Journal* had left the way open for a new investor, Scottish Radio Holdings, which began a process of reinvigorating the management team and refinancing the entire operation. Additional capital of £2 million was raised in November 1997, and, in January 1998, the station was relaunched as Today FM. Canwest, a major Canadian media company, was added to the list of investors. The turnaround was dramatic. Today FM (albeit from a low base) was the only radio station not to lose listeners in the first three months of 1998. In a tribute to its success, RTÉ lured one of its most popular music presenters, John Kelly, to the state station; another, the current-affairs presenter Eamon Dunphy (whose career with the station was temporarily in jeopardy at the time of the relaunch), went from strength to strength, with a daily audience of 144,000 to RTÉ's equivalent of 200,000. At the end of 1999, the radio audience figures disclosed that, for the first time, RTÉ's share of the total radio audience (including the regional stations) had fallen below 50 per cent (to 49 per cent). It was ten years since the first legal broadcasts by commercial stations. The station reported an operating profit of £100,000 for 1999, and now had some 7 per cent of the national audience (5 per cent in Dublin) (*SBP*, 27 Feb. 2000; *ST*, 27 Feb. 2000). It was, ironically, almost exactly the same figure on which Century had found itself incapable of surviving.

One anomaly was that Today FM was not a member of IBI – Independent Broadcasters of Ireland – the successor organization to AIRS (Association of Independent Radio Stations), which had been set up in 1990 as a lobby group, particularly on issues relating to the licence fee and to the transition from analogue broadcasting to a digital environment. Also outside any formal structure were the pirate stations: despite repeated legislative initiatives designed to curb their activities, there were still, in mid-June 2000, at least thirty-two such stations operating in the greater Dublin area, offering fare that varied from religious broadcasting to dance music. Unusually, there were no viable student radio stations. An early experiment, DCU FM, ran from 1995–7, when it ceased broadcasting and its licence lapsed.

The launch of TV3, after a series of tribulations that made those experienced by national commercial radio appear relatively trivial, took place on 21 September 1998. It made steady, if unspectacular progress, dogged to some extent by technical issues. (It was having arguments with Cablelink about its position in relation to other programmes on the cable 'dial', complaining that as an Irish station viewers should be able to access it before imported channels.) It was the first station to start breakfast TV (on 20 September 1999), although audience results were initially disappointing, partly for the technical reasons already mentioned. In November 1999, it secured rights to an important series of football matches for three years from September 2000. By the end of 1999 it had a share of 8 per cent of its main target audience of 16–44 year-olds across all channels (a 16 per cent share of the audience for Irish channels). It had reduced its average monthly losses from £565,000 at start-up to £383,000, and had taken in approximately £13 million in adver-

tising revenue, compared to RTÉ's £121 million. It was also taking a legal
initiative about RTÉ's access to licence-fee income, launching a case in the
European courts alleging that it was the subject of discrimination on this
score (*ST*, 30 Jan. 2000). This is dealt with more fully in the following chapter.
Perhaps more significantly, it had secured investment from Canwest Global,
which already had a 29.9 per cent stake in UTV.

Cross-border activity in the media – particularly electronic media – was
already beginning to adopt new modalities, quite apart from questions of
ownership and control. In July 2000, RTÉ began a marketing campaign to
recruit viewers in Northern Ireland. The problem, as far as RTÉ was
concerned, was that only 19 per cent of Northern homes had cable, a much
smaller proportion than in the Republic. Nonetheless, some 50 per cent of
Northern households had adjusted their sets to receive the RTÉ signal, and
RTÉ aimed to expand that to 75 per cent by targeting retailers and others
with details of the appropriate tuning procedures.

The five-year period up to 2000, therefore, saw a number of radical
changes in the prospects for media in Ireland, not only in broadcasting but
also in print. A combination of technological change, the pressures of globalization and rationalization, and the increasing involvement of non-national
and multinational interests in every sector, was reshaping the environment
and work practices as never before (Hazelkorn, 1996). In the process, new
links were emerging between media in Northern Ireland and the Republic,
even if the shape of future strategic and other alliances was as yet indistinct.
Within the next two years, Ireland would have access to some 200 broadcast
channels, some 98 per cent of them originating outside the state. Control over
the gateways and relay mechanisms through which these channels were
distributed to the Irish population was already, as in other small countries, a
matter of considerable concern, and the cultural and political questions
raised by these developments would become even more pressing in the years
ahead.

9 From uncertainty to boom, 2000–7

By the turn of the century – generally taken to mean January 2000 – the uncertainty that had affected all Irish media since the collapse of the Irish Press group five years earlier had begun to wane. The fragility of the recovery, however, was shortly to be underlined by the dot-com bubble after 1996, and its short-lived collapse in 2000, when all the economic indicators, which had looked set fair, had begun to seem more provisional.

The climate was becoming, if anything, febrile. Takeovers, mergers and acquisitions were in the air, particular in the regional papers and in local radio. The daily papers' readership figures continued to advance healthily, although those for the Sunday titles – a straw in the wind, perhaps – did not. In both print and broadcasting, the move towards digitization, and towards profit-taking (particularly towards the end of the 2000–7 period), led to an increasing number of mergers and acquisitions, and to substantial develop-ments in regulatory oversight, across media sectors generally.

In the decade to 2001, the country's growth rate had averaged above 7 per cent annually, the workforce expanded, as did lending to households, and the top rates of income tax were sharply reduced. Other indicators were more ominous. House prices increased by no less than 17 per cent between May 2000 and May 2001. Online spending by advertisers was still negligible, at about £6 million, but the cost of libel awards for media companies was still high. In 2001 the main opposition party, Fine Gael, published proposals for reform of defamation law that echoed – some said inadequately – a report of the Law Reform Commission on this topic of a decade earlier. Judges, whose comments on media reportage were to become more waspish as the new century progressed, were ever alert to the possibilities of contempt of court.

Some content issues that had been raised a few years earlier continued to reverberate, in particular the issue of child abuse, which had been raised in print by journalists such as Alison O'Connor, Veronica Guerin, Jim Cusack and Ger Walsh, but which did not arouse much popular emotion, or desire for change, until it was taken up by television, and in particular by Mary Raftery's 1999 RTÉ television documentary series *States of Fear*. Newspapers were also becoming more likely to criticize each other, even if only indirectly, and some controversies prefigured a more general, and political, concern about allegedly inappropriate newspaper comment that was to lead, as will be described later, to the establishment of a press council and a press ombudsman. One such related to an article in the *Sunday Independent* on 22 October 2000, in which a columnist described that year's Paralympic Games as 'grotesque'. This evoked strong criticism from at least one government minister, and from groups representing the disabled (*IT*, 24 Oct. 2000).

In the years of the boom between 2000 and 2008, daily newspapers' circulation figures held steady in many cases and, in some cases, improved. The largest-selling daily paper, the *Irish Independent*, which was selling 167,567 copies daily in the first quarter of 2000 (the period to which all other circulation figures for this year also refer), at first improved substantially. In February 2004 it began to publish simultaneous tabloid and broadsheet versions of the paper, which by 2004 were selling a total of 181,080 copies daily. By the first quarter of 2005, when the broadsheet version of the paper had been discontinued, this had slipped to 162,582.

There was a mixed pattern of winners and losers among the other titles between 2000–8. The *Irish Times* had improved marginally, by an average of some 600 copies, as had the Irish *Daily Mirror* (106). The biggest winner was the Irish *Daily Star*, which added an average of almost 27,000 copies daily, an increase of more than 25 per cent. Elsewhere in the daily field, the most significant fall was in the sales of the *Irish Examiner*, which saw its circulation decline by more than 8,000 to 54,191. A significant newcomer was the Irish *Daily Mail*, which was launched in February 2006, and – with low pricing and an editorial policy aimed firmly at the middle of the market – was clearly targeting the *Irish Independent*.

The Sunday newspapers exhibited a negative trend during this period, although some figures were still impressive. The *Sunday Tribune* briefly peaked at 80,000, or 8.4 per cent of the Sunday newspaper market, and a readership of some 280,000, under one of a rapid succession of editors, Paddy Murray. It was alone among Irish newspapers in its trenchant criticism of the invasion of Iraq, but its failing fortunes in the increasingly competitive Sunday newspaper market were eventually to take their toll. Its circulation had fallen to 65,000 by 2005, and it was further encumbered by an intensification of the financial problems discussed earlier.

Circulation losses by 2008 were also sustained by the two major Sunday newspapers in the Independent group, the *Sunday Independent* (down 30,500) and the *Sunday World* (down 16,600). *Ireland on Sunday* had been bought by the owners of the *Daily Mail* in September 2006; rebranded as the *Irish Mail on Sunday* it broadly maintained the 125,000 weekly sales of its predecessor. The

News of the World had decreased marginally to 154,000 – its serious troubles were in the future. A sign of the prevailing optimism in some quarters was the purchase in April 2002 by the owners of the *Irish Examiner* of the *Sunday Business Post*, which was on a modest but attractive growth spurt to a circulation of almost 56,000 by 2008. All in all, while some of the figures were troubling, given the substantial economic growth in the period, most of them were still healthy enough to lull newspaper proprietors and editors into a business-as-usual mentality. This was to be rudely shattered after 2008.

The absence of any form of content regulation other than in relation to defamation and official secrecy – both of them governed by legislation dating back almost half a century – was highlighted in a number of controversies. One occurred on 11 February 2005 when Kevin Myers, a prominent columnist in the *Irish Times*, published a column widely interpreted as criticizing the public welfare system, implying that it encouraged young women to have children outside marriage. Within days, a major controversy had erupted and both Myers and his editor immediately apologized publicly (although in the view of some critics inadequately). That was not, however, the end of the affair. Many of Myers' colleagues on the *Irish Times* joined the chorus of disapproval and, little more than a year later, Myers had signed a lucrative contract with the rival *Irish Independent*.

The issue of regulation was bound up with two other issues that evoked political reactions. One of them was defamation, and the other was privacy. Although the Commission on the Newspaper Industry had reported in 1996, and had called for some reforms in both these areas, no legislative initiatives had been taken. The government that took office following the general election of 2002, which comprised both Fianna Fáil and the Progressive Democrats, had included in its agreed programme a commitment to move, within the context of a statutory press council and improved privacy laws, to implement reforms of libel laws designed to bring them into line with those of other states. This commitment resulted in the creation, in 2002, of a legal advisory group, which reported the following year.

While it recommended a number of changes to the law of libel, some of which would have met with the agreement of the newspaper industry, this advisory group also, and more controversially, proposed the establishment of a statutory press council, whose members would be appointed by government, which would articulate a code of conduct for the press, and have very considerable powers. Although these powers should not, it suggested, include the power to award damages, they were nonetheless substantial:

The remedies available should ... include directing the relevant publication to publish the substance of the Council's decision or to publish a correction or retraction of the material complained of. The Council should also have the ability to give directions as to the manner in which a correction, for example, should be published. If a publication were to refuse to comply with a decision of the Council, the Group suggests that the Council should be empowered to apply to the Circuit Court for an order compelling compliance. Failure to comply with the court order could result in the publication in question being found to be in contempt of court. Ideally, were any monetary fine to be imposed for that contempt, such fine would then be remitted to the Council itself. (Legal Advisory Group on Defamation, 2003: 16)

This report had several immediate effects. The first was a tsunami of criticism from the print media and from a range of commentators, warning against what they saw as the threats to free speech. The second was an initiative by the newspaper industry as a whole, under the umbrella of the National Newspapers of Ireland, to set up a press industry steering group, chaired by a former provost of Trinity College Dublin, Professor Thomas Mitchell, and facilitated by the former Northern Ireland ombudsman, Dr Maurice Hayes. It included representatives of all newspapers circulating in the Republic of Ireland, as well as the National Union of Journalists, which had upwards of 3,500 members in the Republic, and which had never participated in the UK's press complaints commission.

This helped to stall government action on the proposals of the legal advisory group, and, although the press industry group operated in an atmosphere of strict confidentiality, it became known at a later stage that informal communications had been maintained with the department of justice in an attempt – successful, as it turned out – to obviate the need for the government to proceed with the establishment of a statutory press council. It did this by devising an alternative model incorporating both a press ombudsman and a press council, which, although independent and non-statutory, would be formally recognized by the government under statute, would have a lay (i.e. non-industry) majority, and would operate on the basis of a code of practice devised by the press itself. This model went a considerable distance towards achieving the government's stated regulatory objectives, and, some two years after it had been established voluntarily by the industry in 2007, was effectively recognised in the 2009 Defamation Act.

A related government initiative was a proposal in 2006 for legislation giving a statutory basis for legal action based on invasions of privacy. This initiative was also in the name of the minister for justice, notwithstanding that a year earlier he had expressed himself as opposed to such a development and supportive, instead, of the idea that the protection of privacy was best served by an organic case-by-case build-up of jurisprudence. The minister, Michael McDowell, defended his 2006 proposal by citing two examples: one in which a newspaper journalist had photographed a fashion model topless while she was in a changing room at a fashion show, and another in which a newspaper had sent a photographer to a hospital in an attempt to secure a picture of a television presenter and her baby just after she had given birth (*IT*, 17 Sept. 2006). The furore that greeted McDowell's proposal, particularly from the media, was so intense that the proposal disappeared from sight and was not resurrected until after 2007, and then ineffectively.

The acquisition of the *Belfast Telegraph* in 2000 marked a high point in the company fortunes of INM and in the personal scenario of its chairman. However, in some respects the future was becoming more problematic for the company, not least because of the high cost of the *Belfast Telegraph* acquisition. Debt at INM had remained at a relatively moderate level even during the aggressive expansion of the 1990s, or at least up to 1997, but, by the end of 2001, the group's net debt had increased to €1.376 billion, a gearing of 140 per cent, while the interest cover ratio (a measure of the capacity to service interest payments of debt) had fallen to 2.3. There had also, inevitably, been a significant increase in interest payments. This added up to a level of indebtedness which served notice that it was time to steady the ship. By the end of 2003, this exercise had been carried out with some verve and to considerable effect. In the course of this restructuring, the group reduced its total net debt by 245 million, or some 20 per cent, to £978 million, of which only £644 million was recourse debt (i.e. backed by INM collateral). At the same time, the measures taken had important structural implications in Australasia and in the UK, as well as involving two separate rights issues and a number of key disposals.

Towards 2007, however, there were ominous indications of problems, particularly in relation to advertising. Part of it related to the second Gulf War, which resulted in a fall-off in media advertising globally. Independent News and Media, possibly reacting to the staff reduction programme initiated by Johnston Press in that year, announced in December that in spite of a sharp slowdown in property advertising, it was forecasting at least single-digit growth for 2008. Events were to prove this forecast more than a little optimistic. And, while the economy has always been the principal driver of the financial health of Irish media, the early noughties also saw the development of technological innovations that were, in the long run, to combine with the economic downturn of 2008–16 to dramatic effect.

During this period, content issues in the print media in Northern Ireland had, since the Good Friday Agreement of 1998, largely evaporated. Nonetheless, academic studies began to appear, featuring in particular discussions about the role that journalism had played in the Northern Ireland conflict generally and, more particularly, in the period up to 2005. These ranged from the broadly supportive (Kovarik, 2006), through the inconclusive (Fawcett, 2002; Baker, 2005), to studies that were unambiguously sceptical about the power, or causal effectiveness, of such journalism (Hanitzsch, 2004, 2007; Hackett, 2006). One case study of two newspapers on opposite sides of the nationalist/unionist divide in Northern Ireland, indeed, underlined the stereotyping and oversimplification displayed by traditional news reporting that had 'unreasonably accentuated and heightened' the polemic, and the resultant physical confrontations, between citizens of differing political views (Peleg, 2006: 7). A more positive analysis, specifically of television, suggested that this medium had:

> played a central role in political exchanges through the peace process, as political participants routinely used it as a communications conduit … to convey positions, responses and proposals as part of a process of dialogue, accelerating diplomatic communication and moving information in the direction of groups beyond the normal reach of government. (Corcoran, 2004: 123)

In the technological area generally, both in Northern Ireland and in the Republic, the internet and its applications were not new in 2000, but newspapers' adaptation to this phenomenon, both in terms of web publication and journalism practice, was initially slow. Nonetheless, newspapers were quick to organize web presences for themselves – a poll carried out in June 2000 found that news outlets were accessed by 64 per cent of users, ranked second highest behind search engines (Amárach, 2000). Another survey the following year found that five of the top sites for page impressions were those hosted by established media, including the *Irish Times*, which with over 24 million hits. a month (many of them from internet users abroad) was by far the most popular; the *Irish Independent* website had, at that stage, only 5.6 million hits per month. By April 2002 the percentage of households with internet access was still comparatively low, at 32 per cent (ODTR, 2002a). Access and usage was to grow exponentially in the years ahead.

Journalism itself was relatively slow to adapt to the new technology. One comparative study of four countries, including Ireland, disclosed that inter-activity was low on the main Irish site, Ireland.com, owned by the *Irish Times*. 'Although user feedback has played an important role in Ireland.com's devel-opment', the study noted, 'it has not been a primary factor in shaping it.' They concluded:

> The majority of the sites surveyed implement a traditional news production model, reinforcing the role of the editor, sub-editor and other established production roles in online news provision. Most of those implementing a more user-orientated strategy see this strategy as an extension of the model of professional journalism rather than a reinvention. They are not attempting to rewrite the rules of journalism, they are simply allowing the relationship between the user and the producer to become more active and allowing this to inform and reinvig-orate the content carried on the site. (Trench and Quinn, 2003: 7)

As was the case in the press sector, although the economic growth associated with the Celtic Tiger of the mid-2000s saw both state and privately owned broadcasters in apparent financial good health, the opening years of the decade were significantly more turbulent for broadcasters, especially for RTÉ. While the initial post-1988 wave of competition had already trans-formed the broadcast firmament, the impact had been largely limited to the medium of radio. After 2000 new television-distribution platforms brought an exponential increase of overseas channels and even relatively modestly capitalized indigenous enterprises perceived opportunities to establish new television stations. Their presence changed the de facto rules of the Irish broadcasting market, rules which would also see de jure changes as the European Union came to exert a new influence, in sometimes unexpected corners.

RTÉ entered the new millennium facing structural difficulties. The expansion of the 1970s and 1980s had left the station with a substantial workforce relative to its output. By contrast with leaner commercial and public-service rivals, in the first case (TV3) untrammelled with heavy public-service obligation and in the second (TG4) unburdened with production activities (TG4 was effectively a publisher-broadcaster), RTÉ looked sluggish. Associated personnel costs were exacerbated by national wage agreements negotiated under the social partnership model of the 1990s and 2000s.

These costs were not compensated for by revenue increases. A 1998 financial surplus was followed by deficits of £16.6 million in 1999 and £11.2 million in 2000. The £8 increase in the licence fee in 1996 (itself the first since 1986) was largely absorbed by RTÉ's annual obligation to provide – at no charge – 365 hours of content to TG4. RTÉ's monopoly of television advertising was, by 2000, challenged by new local channels and UK-based broadcasters.

The scale of competition increased after 2000 – in February 2001 the IRTC awarded the Cork youth licence to the Thomas Crosbie-backed RED Hot FM while in the same month the supreme court upheld an earlier IRTC decision to award the Dublin youth licence to the Denis O'Brien-backed Spin FM, following a challenge from a rival bidder. The Dublin market was further augmented by the launch of another O'Brien-backed station, Newstalk 106, in April 2002. Even if the new market entrants did not necessarily immediately thrive (Newstalk, in particular, found it hard to move beyond a 2 per cent share of the Dublin market in its early years), they ate away at the market share and commercial revenues of incumbents.

Though RTÉ still accounted for the most successful individual radio shows, in 1999 its share of the national radio market fell below 50 per cent for the first time, as the combined weekday listenership of Today FM and local stations reached 52 per cent. Even the launch of a fourth RTÉ station – the classical-music-focused station Lyric FM – in May 1999 could not stop RTÉ's share further declining to less than 40 per cent by 2006. Furthermore, as new stations tended to focus on younger (15- to 34-year-olds) – and therefore more valuable from a marketing perspective – audiences, the market share of the new entrants tended to understate their share of advertising revenues.

Competition in television advertising was similarly complicated. The launch of Sky Digital in 1998 had ushered in a wave of new channels. NTL's acquisition of Cablelink in 1999 should have quickly expanded the number of channels available to their subscribers too, but post-dot-com bubble funding difficulties delayed this somewhat. Nonetheless, from 1998 a new 'opt-out' innovation allowing UK channels to target ads at Irish audiences hugely expanded competition for Irish television ad revenue. By early 2003, Irish advertisers could choose from five channels to advertise on beyond RTÉ1, Network 2, and TG4: TV3, NTL's local cable channel, Sky One, Sky News and E4. Additional UK opt-outs saw seventeen channels chasing Irish television advertising revenues by the end of 2005.

The most overt challenge to RTÉ's dominance appeared almost overnight in September 2000 when Granada Television, one of the two largest ITV franchise holders in the UK, acquired a 45 per cent stake in TV3, putting it on parity with the other major shareholder, Canadian giant Canwest (which itself had acquired a 29.9 per cent of UTV in 1997). Although TV3's position had improved since its launch, with increased weekly broadcast hours and reduced monthly losses (though these were still £400,000 a month by January 2000), its 7 per cent share of peak-time viewing trailed far behind the 41 per cent cumulatively enjoyed by RTÉ1 and Network 2. By April 2000, just 29,000 viewers watched the channel's 7 p.m. bulletin compared with the nearly 400,000 tuning in for RTÉ's *Six One News.* The channel's position was not helped by its reliance on imports and cheaply produced studio-based local content, which belied Paul McGuinness' 1989 claim that TV3 would treat the local independent production sector as a major source of content.

Nonetheless it was another imported show – Granada's *Coronation Street* – that not only helped boost TV3's ratings to a point where the channel could break even, but which prompted the Granada takeover in the first place. '*Corrie*' had been a key element in RTÉ's midweek schedule since 1977, and TV3 had made several unsuccessful approaches to Granada in an attempt to poach it. However, by 2000, an expansionary Granada regarded TV3 as a means of entering the neighbouring market which they could treat as akin to another UK region.

Thus, in 2000, Granada responded to another TV3 bid for *Coronation Street* with a proposal to take an equity stake of £38.05 million in the Irish broadcaster. Though initially looked at askance by the IRTC, a TV3 assurance that the investment would mean more independent production won the regulator over. *Coronation Street* immediately increased TV3's ratings in its timeslot tenfold: 600,000 people (nearly twice TV3's previous peak audience) watched the first episode on TV3 in January 2001 (compared with the 160,000 watching RTÉ's replacement show). Within 12 months TV3's overall prime-time share had risen four points to 11 per cent while RTÉ1's dropped three to 28 per cent.

The episode revealed much about the nature of competition within the Irish market: audiences were secured by offering the familiar rather than the innovative. 'Simulcasts', whereby ITV content was simultaneously broadcast on TV3, came to constitute up to 20 per cent of the latter's prime-time schedule, especially after the rise of 'shiny floor' shows like *The X Factor* after 2004. Mimetic strategies were not exclusive to commercial broadcasters: although 2000–7 saw a marked increase in domestic dramas on RTÉ (*Bachelors Walk, The Clinic, Showbands, On Home Ground, Pure Mule, Love Is the Drug* and *Legend*), RTÉ responded to losing *Coronation Street* by acquiring another UK soap – the BBC's *Eastenders* – having dropped the home-produced rural soap *Glenroe* in January 2001. This couldn't compensate for losing 'Corrie'. While the television advertising market as a whole grew 3–4 per cent between 2000 and 2002, RTÉ's commercial revenues fell 10 per cent in the same period.

An over-reliance on advertising revenues not only left RTÉ vulnerable to the vagaries of the market but also arguably undermined RTÉ's public-service status. By 2000, just €84 million (or 32 per cent) of RTÉ's total income of €264 million was generated by the licence fee, leaving it more dependent on commercial revenue than any other public-service broadcaster in Western Europe. This created an incentive for the state-owned broadcaster to show content that targeted advertiser-friendly markets, to the potential detriment of programming for other, more niche, audiences.

RTÉ's response to the unfolding financial crisis was shaped from the top down. In 2000, Cathal Goan replaced Joe Mulholland as director of television. As an old hand at RTÉ, Goan was noted for his commitment to locally produced content, and his track record in running TG4 on a shoestring certainly helped his candidacy. The new RTÉ Authority appointed in May 2000 included a marked emphasis on a commercial outlook. Incoming chairman Paddy Wright, a former group president at Jefferson Smurfit, described RTÉ's costs as 'out of control' (*IT*, 13 June 2000) and expressed a desire for the station to become more commercially focused.

Cost-reduction and commercial-revenue enhancement constituted two elements of RTÉ's response. In 2000, the broadcaster sought 300 voluntary redundancies and an overhaul of work practices from remaining staff. Though designed to reduce current expenditure by €15 million per annum, in the short-term these measures worsened the situation as the exceptional expenditures incurred drove the overall deficit to €70.9 million in 2001.

RTÉ increasingly looked to sponsorship deals to plug the commercial-revenue gaps: even *The Late Late Show* succumbed to a deal with Renault. In May 2001 concern at some ethically questionable relationships – an insurance company sponsoring a domestic-crime show and a bank supporting the station's flagship finance programme – saw the introduction of some new internal guidelines prohibiting the sponsorship of current-affairs shows. Nonetheless, as broadcast advertising was increasingly cannibalized by other media, sponsorship would remain a crucial component of the new commercial dispensation at RTÉ: by September 2007, there were twenty-three such programmes in the RTÉ schedule.

The other key plank of RTÉ's financial strategy rested on the licence fee. By 2000, the value of the licence was limited not only by the small scale of the Irish market but by the fact that, at £70, it was lower than in any other EU member state, barring Italy and Greece. (The situation was exacerbated by the fact that An Post retained roughly 10 per cent of the revenues as a collection fee, and 10–15 per cent of the population simply evaded the charge.) In October 2000, the station formally applied for a significant increase in the fee – to £120 – while asking that future increases be indexed to inflation, rather than left to ministerial discretion.

The minister for arts, culture and heritage, Síle de Valera, insisted that any increase depended on an objective evaluation of how efficiently RTÉ was using public funds. TV3's arrival in 1998 in particular had radically altered the tone of the public discourse around RTÉ's dual access to licence-fee funding and advertising revenue. The commercial station repeatedly complained that the absence of a clearly defined public-service remit governing RTÉ's use of the licence fee allowed it to unfairly compete with commercial broadcasters both by offering artificially low advertising rates and outbidding commercial competitors for imported content. Notwithstanding various protocols to the EU treaties that permitted state funding to public-service broadcasters, TV3 asserted that RTÉ's access to the licence fee contravened EU competition provisions, and filed a complaint to this effect with the European Commission in 1999.

It would take nearly a decade for the impact of that complaint to be fully felt, but in March 2000 Síle de Valera noted that a new Broadcasting Act would clarify the public-service remit, implicitly determining how RTÉ could spend the licence fee thereafter. On its passage in March 2001, the Act included the first overt legislative description of the state broadcaster's remit, echoing the traditional Reithian emphasis on information, education and entertainment. (The act also prepared for a decoupling of RTÉ from both TG4 and – in anticipation of a shift to digital broadcasting – from its function as a broadcast transmission network provider.)

Although for TV3 this description offered insufficient clarity on exactly what could and could not be funded using licence-fee revenues, RTÉ recognized the primacy of a new discourse of transparency. In May 2001 it made the symbolic gesture of agreeing to make public the salaries earned by its leading broadcasters. Addressing an Oireachtas committee, RTÉ Director General Bob Collins accepted the need to be more formally accountable: 'Proof is required: facts and figures that nobody can distrust. We have to be accountable' (*IT*, 30 May 2001).

However, such rhetorical commitments – together with a June 2001 announcement of further expenditure cutbacks of £7 million – did not secure RTÉ the sought-for licence-fee increase. In July 2001, based on advice from PriceWaterhouseCoopers (PWC), de Valera announced an increase of just £14.50, less than a third of what RTÉ had sought. PWC recommended that funding for new programmes and the digital transition should be withheld until RTÉ could demonstrate that it could reduce and control its costs. In addition, de Valera argued that she could not countenance increasing the flow of public monies to RTÉ when it was sitting on extensive – £145 million – cash reserves earned from the sale of Cablelink:

> I could not stand over a situation where the licence fee payer was required to fund balances of the order maintained by RTÉ at present. Spending some of these exceptional cash reserves on maintaining the quality of existing services could not be deemed to be 'frittering away' resources. (*IT*, 10 June 2001)

For some, the rejection of the application evinced the state's (and more specifically Fianna Fáil's) ongoing antipathy towards the broadcaster, especially given its coinciding with a government plan to privatize RTÉ's transmission infrastructure and the retailing of digital multiplexes. Critics of RTÉ countered by asking whether a channel already reliant on US and UK imports could really lay claim to the mantle of public-service broadcaster, and the funding associated with that status in the first place.

Publicly, RTÉ took the decision on the chin, acknowledging that PWC had made 'legitimate observations', while committing to assemble a more robust application for a subsequent increase. Recognizing that the language of management consultancy increasingly held sway in such negotiations, RTÉ hired its own consultants to conduct a strategic review. Reflecting the crisis mode within RTÉ, KPMG were permitted to contemplate a scenario whereby RTÉ ceased all internal production, bar news and current affairs, becoming

effectively a publisher-broadcaster and entirely selling the Montrose facility –
although this option was not pursued. The consultants emphasized the need
to overhaul the station's internal accounting practices to separate out how
different sections of RTÉ were performing and thus to achieve the kind of
financial transparency PWC had found lacking. Bob Collins had publicly
committed to addressing this in November 2001 and from 2002 the station's
accounts were reorganized around six new integrated business divisions:
radio, news, television, network, publishing and performing groups. In
theory, at least, the adoption of these structures made it feasible to distinguish how and where licence-fee revenue and commercial revenues were spent,
a distinction reflected in RTÉ's annual reports from 2003 onwards.

   In the interim, however, RTÉ continued to experience something of an
existential crisis. In November 2001 the station announced the need for
further cuts of £24 million, axing another 150 jobs (on top of the 330
negotiated the previous year). An immediate reduction of £6.9 million in
programme-production budgets accounted for the single largest cutback,
leading to a 200-hour reduction in internal output. However, RTÉ also
divested core facilities such as their Outside Broadcast operations
(contracting out that function thereafter). A further £1.3 million was cut from
the news budget leading to the closure of Lyric FM's entire news service.
Even this was not enough. By May 2002, chairman Paddy Wright was moved
to announce that the station was no longer merely facing a financial deficit
but its 'death-knell' (*IT,* 22 May 2002). Three months later in July 2002, the
station scrambled to find *another* 200 redundancies, and further scaled back its
commissioning of programmes.

   The problems in RTÉ's finances coincided with a second broadcasting-
related crisis for the government, connected to the introduction of digital
terrestrial television (DTT). In January 2000, RTÉ was planning to launch
three digital channels on that platform. By March, these plans had been
abandoned due in part to the station's financial difficulties, and in part to
confusion over the ownership and control of the new digital transmission
network.

   In October 2000, anticipating the 2001 Broadcasting Act's requirement to
sell a majority share of their transmission network to a private operator, RTÉ
had appointed the London investment bank Rothschild to handle the sale.
Síle de Valera's intention was that the transmission business would provide its
services to a new, separate and privately owned digital TV multiplex operator,
which would negotiate with broadcasters to offer television channels (and

potentially other data services) directly to consumers, in competition with existing cable operators. The established national broadcasters – RTÉ, TG4 and TV3 – were guaranteed access to these multiplexes. For consumers, this promised to bring some of the fruits of the information age to those who lived beyond the reach of cable infrastructures and who did not wish to subscribe to Sky's satellite services.

However, the political delay between the initial publication of plans for a digital upgrade of the terrestrial television-distribution system in 1997 and the passage of the 2001 Broadcasting Act had seen other distribution entities acquire first-mover advantage. It initially appeared that cable companies might be the primary beneficiaries of the DTT delay. Having acquired Cablelink in 1999, NTL had committed to a network upgrade, while Independent News and Media's Chorus was reportedly spending £300 million in 2000 to upgrade its cable/MMDS services for a digital offering. Even Eircom, anticipating NTL's and Chorus' entries to the voice telephony market, began upgrading its own ADSL system to facilitate the delivery of digital services (internet, email and video on demand) via its infrastructure.

Digital promised a different relationship with television. Rather than the relatively passive experience of analogue television, digital promised an interactive experience, not least because the multiplicity of channels it made possible could only be navigated via electronic programme guides (EPGs). Furthermore, digital *cable* infrastructures, in particular, promised something neither DTT or satellite could compete with: a future in which Irish internet users would be liberated from a reliance on dial-up modems with slow download speeds by the advent of 'always on' broadband services bundled with TV offerings and voice telephony.

However, in January 2001, NTL informed ComReg, the Irish communications regulator, that it could no longer meet the digital rollout terms built into its licences. Having spent €680 million acquiring Cablelink in 1999, NTL's parent company was now on the verge of bankruptcy. For its part, Chorus recorded losses of €36.6 million in 2001 curtailing its spending on network upgrades. Fewer than 20,000 homes received digital-television services via cable by 2001, allowing Sky Digital, launched in 1998, to enjoy first-mover advantage, securing 193,000 Irish digital subscribers by November 2001.

DTT's difficulties continued to mount. Although the Broadcasting Commission of Ireland (as the Independent Radio and Television Commission had been renamed after the 2001 Broadcasting Act) had moved quickly to licence digital-multiplex operations in 2001, only one group had actually lodged a bid by the autumn 2001 deadline, and even that was withdrawn in October 2002. The lack of interest may have been influenced by the fact that Sky had already secured 20 per cent of the potential market, but the contemporaneous difficulties of ITV's ONDigital in the UK (which would cease trading in June 2002) can hardly have been encouraging. This had implications for the sale of RTÉ's infrastructure: if there was no market for DTT, then RTÉ's transmission network became less valuable, especially as Sky continued to vacuum up homes that might otherwise have become DTT subscribers.

Indeed, RTÉ's own doubts about the future of DTT were such that in November 2001 it negotiated for a place on the Sky Digital platform. Though damaging for the prospects of DTT, RTÉ eyed Sky's ability to secure nearly 200,000 Irish subscribers without the carrot of RTÉ programmes somewhat nervously. The possibility of a substantial (25 per cent) cohort of Irish homes without RTÉ undermined the station's claims to be both a public-service broadcaster and to be capable of delivering advertising to a wide audience. For its part, Sky regarded the ability to offer RTÉ as part of its basic service as crucial to its longer-term expansion ambitions. Its Irish subscriber numbers rapidly climbed to 272,000 by March 2003 and in May 2005 it surpassed the cable companies to become the largest pay-TV operator in Ireland, with 355,000 subscribers.

As the minister in charge when the RTÉ funding crisis and the DTT saga unfurled, Síle de Valera received criticism from political opponents, but also in the broader public sphere. The limited scale of the July 2001 licence-fee increase and the apparent intention to entirely divest RTÉ of its transmission function were cited as further evidence of a political desire to hobble RTÉ. In a move clearly intended to defuse, or at least deflect, such criticism, in March 2002 de Valera announced the establishment of an independent forum on broadcasting. In a tacit acknowledgement of the crisis faced by RTÉ, the forum's remit focused on the future of public-service broadcasting, although notably, its original terms of reference prohibited it from referring to the adequacy of the licence fee.

The various submissions to the forum clearly reflected the interests of those making them. RTÉ continued to insist on an understanding of public-service broadcasting which required that the 'full gamut' of audience interests were provided for by the public-service broadcaster. It also argued that licence-fee reviews should be conducted independently of both RTÉ and the state. The BCI advocated the introduction of a single regulatory system overseeing both RTÉ and independent broadcasters. By contrast with RTÉ, submissions from commercial broadcasters reflected a perspective character-istic of an increasingly marketized broadcasting landscape that regarded the public funding of public-service broadcasters as a disturbance of normal market conditions. Thus, commercial broadcasters tended to propose that, in a competitive market, the function of public-service broadcasting should be limited to addressing market failure, producing only those programmes that the commercial sector would not.

However, the submissions from commercial players also introduced the possibility that public-service content was no longer synonymous with public-service broadcasters. Irish audiences, it was suggested, were indifferent as to what channel the programmes they watched were on, as long as they included some high-quality, local content. To that end, their submissions suggested replacing the licence fee with a levy on subscriptions to television platforms. This would fund the production of programming that met specified public-service criteria and which, as the Sky and Today FM submis-sions made clear, could be availed of by broadcasters of any stripe, public or otherwise.

The May 2002 general election saw a reallocation of responsibility for broadcasting back to the department of communications, and thus the recommendations of the forum, published in August, were addressed to that minister, Dermot Ahern, who had stated on taking office that addressing RTÉ's problems would be his first priority.

The forum's conclusions offered, in effect, 'something for everyone in the audience'. In line with the private sector's complaints about the opaque nature of RTÉ's remit, the forum argued that RTÉ's performance should be measured against a detailed statement of public-service broadcasting commit-ments encapsulated in a charter agreed with the relevant minister. The charter would include clear criteria relating to content, impartiality and the appli-cation of public sector funding, under which RTÉ's performance could be assessed by the broadcast regulator. RTÉ was also to publish transparent accounts for each service it offered and, specifically, for its use of public funding.

Having been given informal but specific permission to do so by Ahern, the forum also made some comments on the licence fee. Noting that the RTÉ public-service charter would establish a base cost for the public-service remit, the forum recommended that subsequent changes in the level of the licence fee should be index-linked to inflation – i.e. removed from the discretion of politicians – subject to RTÉ's meeting its charter commitments.

The forum also identified, as a matter of urgency, the need to revisit the DTT issue, and called for the creation of a new single broadcasting regulator, to be called the Broadcasting Authority of Ireland (BAI). The latter would assume the functions of the Broadcasting Commission of Ireland (BCI), and the regulatory functions of the RTÉ Authority. In a none-too-subtle critique of the commercial sector's sometimes patchy track record in meeting its licence commitments (and the light-touch approach to their regulation), the forum also emphasized that the new regulator should actively seek to stimulate *meaningful competition* in the commercial sector, and to ensure broadcasters were actually addressing the needs of the communities they served. It further recommended that the BAI should ensure that commissioning of programming from the independent production sector would become a precondition for the granting of broadcasting contracts.

Unsurprisingly, the commercial sector strongly welcomed the suggestion that RTÉ be made more accountable for its public-service remit. For its part, RTÉ welcomed the measures to index-link licence-fee increases to inflation. The government implemented virtually every recommendation the forum made, although the time scales for doing so varied. Though Dermot Ahern announced plans to legislate for the creation of a BAI in December 2003 it was September 2009 before it actually took office. By contrast, it took just four months for him to grant RTÉ an additional €36 increase in the licence fee, in return for new programming and reporting commitments. Anticipating complaints from the private sector about the increase – which duly came – RTÉ were careful to stress that the additional revenue would be spent on a narrow – but politically safe – news- and current-affairs-oriented definition of public service rather than light-entertainment shows.

In May 2003, RTÉ formalized its public-service commitments in a statement to Dermot Ahern. This emphasized the production of additional local content, specifically committing RTÉ to quantified increases in drama, news and current affairs, music, factual and entertainment. At the same time, it became clear that – index-linking of the licence fee notwithstanding – the state would not entirely abandon its discretionary power over RTÉ's funding. Given that inflation was running at 2.9 per cent in 2003, the licence fee should have increased by around €3.50. However, in November 2003, again informed by a PWC consultancy review, Ahern limited this to €2, noting that although RTÉ had achieved greater transparency in its accounting practices, there remained insufficient evidence of reduced staffing. This established a precedent: though Ahern's successor as minister for communications, Noel Dempsey, would grant three further increases between December 2004 and November 2007, all were significantly below the prevailing level of inflation.

The benefits of licence-fee increases, in any case, no longer flowed only to RTÉ. As part of the post-forum measures, Dermot Ahern had announced that €7 of the licence fee (later set in the 2003 Broadcasting (Funding) Act at 5 per cent of total licence-fee revenues) would be set aside to finance a BCI-administered scheme to be known as the Sound and Vision fund. This was a clear response to calls from commercial players for a competitive fund supporting public-service content production. The forum itself had rejected such a fund, suggesting that it would simply fragment the public-service element of broadcasting. Dermot Ahern's own officials had also disagreed with the move on the grounds that 'once the principle has been established that private broadcasters can benefit from the process of the licence fee, then the Irish broadcasting landscape will have changed dramatically' (*IT*, 4 Feb. 2003).

Against this, Ahern noted that despite an 'explosion in the number of channels on offer to viewers … this has not resulted in an equivalent increase in the choice of programming available to viewers'. Furthermore, he expressed concern that further competition would encourage all Irish broad-casters (public and private) to 'move towards a generic schedule in pursuit of the most economically advantageous audiences' (SD, 4 Nov. 2003). The creation of the new fund was designed to 'incentivize and encourage the production of more innovative and better indigenous programming' (*IT*, 4 Feb. 2003). In practice, the scheme was informed by a narrow conception of public service, which, while emphasizing the need for quality content, seemed primarily informed by 'an identity-driven conception of public service content' (Flynn, 2015). Regardless, the introduction (and subsequent expansion) of the scheme clearly signalled that the equivalence between public-service broadcasting and public-service *broadcasters* prevailed no more.

The forum's emphasis on a tighter definition of RTÉ's public-service remit was shaped by increasingly vocal pronouncements from the European Commission in this area. In September 2001, a month after the forum's interim report, the commission published proposals on the public funding of broadcasters, requiring member states to provide clear and precise definitions of their public-service remits in specific service contracts. The commission had long argued that assessment of the proportionality of state funding for public-service broadcasting required a clear definition and typology of public-service obligations. This reflected the logic of a protocol appended to the 1997 Treaty of Amsterdam that permitted state funding of public-service broadcasting 'insofar as such funding is granted to broadcasting organizations for the fulfilment of the public-service remit *as conferred, defined and organised* by each Member State [emphasis added]'.

This approach was given formal expression in the commission's November 2001 'Communication on the Application of State Aid Rules to Public Service Broadcasting', which pressured member states to arrive at mechanisms that could capture such definitions. The RTÉ charters from 2003 did not meet these standards, and in March 2005, on foot of the original TV3 complaint from 1999, the EU competition commissioner, Neelie Kroes, wrote to the Irish government seeking a more precise legal definition of RTÉ's public-service remit. While not suggesting that RTÉ was in receipt of excessive public funding, the commissioner stressed the need for transparency to ensure that public funding of RTÉ was limited to the minimum needed to achieve its public-service remit.

It was only with the passage of the 2009 Broadcasting Act that the commission's concerns were addressed. Nonetheless, the EU's influence was definitively felt in other aspects of Irish broadcasting, illustrating the extent to which the Irish state could no longer single-handedly dictate broadcasting policy. The regulation of television distribution was a key example of this. NTL and Chorus complained that while they were subject to the oversight of (and a levy by) ComReg, the Irish telecom regulator, their main competitor, Sky, fell beyond its reach by virtue of the fact that it was headquartered in the UK.

From late 2002, however, ComReg looked to extend its purview to Sky by exploiting the forthcoming EU Electronic Communications Framework directive, which was designed to harmonize the conditions under which European providers of electronic infrastructures operated. ComReg's assertion — strongly refuted by Sky — was that regulatory equivalence with NTL and Chorus could only be harmonized by bringing the satellite station within ComReg's regulatory purview. In parallel with this, in 2003 Dermot Ahern made a submission to an EU consultation on revising the Television Without Frontiers directive, arguing that broadcasters should come under the regulatory jurisdiction of the member states into which they targeted television channels, even if they were headquartered elsewhere. This challenged a lynchpin of EU broadcasting policy embodied in the 1989 Television Without Frontiers directive; arguing that a multiplicity of inconsistent national regulatory regimes would discourage pan-national broadcasting, the directive insisted that broadcasters were subject only to regulation in the country in which they were headquartered. Nonetheless, Ahern announced his intention to use Ireland's January-to-June 2004 EU presidency to change this element of the directive if necessary.

Part of the context for this were BCI efforts in 2003 to introduce stricter codes for advertising relating to children. The presence of UK broadcasters offering opt-out advertising on Irish television platforms raised questions about the efficacy of such codes, given that they only applied to RTÉ, TG4 and TV3. For their part, Irish broadcasters complained that introducing such regulations would simply encourage Irish advertisers to spend overseas, doing little to protect Irish audiences while harming the finances of Irish broadcasters. And, significantly, opt-outs were not necessarily limited to advertising. In May 2004, Sky launched *Sky News Ireland*, two daily 30-minute news bulletins exclusively available in the Irish market, fronted by former TG4 and TV3 presenter Gráinne Seoige. Although persistently low audiences saw the service pulled in November 2006, it remained a frustration for Irish regulators that such content lay beyond their reach.

The efforts of ComReg and Ahern, however, came to naught. By June 2003 it became apparent that the Electronic Communications Framework directive, far from expanding ComReg's reach, curtailed it. Regulatory harmony was achieved by reducing the regulatory scrutiny to which NTL and Chorus were subject. Furthermore, the European Commission strongly resisted any change to the principle that a broadcaster's originating country was responsible for its regulation and licensing.

However, if the EU limited the scope for an independent Irish broadcasting policy, it also emerged as a defender of the Irish public interest with regard to the broadcasting controversy attracting the most public attention in the 2000s: the acquisition by Sky Television of exclusive rights to televise the Republic of Ireland soccer team's matches in summer 2002.

As the TV3/*Coronation Street* narrative recounted above demonstrated, competition for audiences meant competition for the rights to popular programming. In April 2000, the Irish independent production company Tyrone acquired the Irish rights to the hugely successful *Who Wants to Be a Millionaire?* quiz format from the UK production company Celador. In August, Tyrone announced that RTÉ would screen the show, with veteran

broadcaster Gay Byrne as the host. Given that the RTÉ version had to
precisely conform to the show's original UK format, the decision leant weight
to the criticism that RTÉ was increasingly driven by commercial rather than
creative considerations. Nonetheless, the first episode drew an audience of
810,000 (half of all Irish viewers), which appeared to demonstrate the –
commercial – validity of RTÉ's decision.

The most intense inter-broadcaster competition, however, was reserved for
those sporting events that retained a capacity for guaranteeing remarkable
television audiences (and thus advertising revenues). GAA games were at the
top of the pyramid (in 2004 the all-Ireland football and hurling finals
averaged audiences of 744,000 and 705,000, respectively, for RTÉ), but the
increasing popularity of rugby in Ireland over the 2000s saw Ireland-England
clashes in the Six Nations tournament routinely winning average audiences in
excess of half a million viewers. Inevitably, new market entrants sought access
to attractive fixtures. By 2001, TV3 had already acquired exclusive Irish rights
to both UEFA Cup and Champions League coverage. In March 2001, Sky
offered the FAI £4 million for five years of the UK broadcast rights to
Republic of Ireland home internationals. The possibility that such deals
might prevent Irish viewers from watching events featuring the national team
had already been demonstrated between 1998 and 2002 when the English
Rugby Football Union had sold Sky exclusive broadcast rights for England's
home games in what was then the Five Nations tournament. As a consequence, only Irish viewers willing to pay a Sky subscription could watch
Ireland-England games played at Twickenham in that period.

Anticipating this possibility, in 1997 the European Commission had
amended the 'Television Without Frontiers' directive to permit member states
to draw up lists of events that had to be made available via free-to-air broadcasters. Yet by 2002, despite negotiations with the FAI, IRFU and GAA, the
department of arts had been unable to come up with a list satisfactory to all
the parties concerned.

Even though the previous decade had made manifest the ambitions of
broadcasters to acquire such rights, it apparently came as a shock to the Irish
government when, on 5 July 2002, the FAI announced the sale to Sky of four
years of exclusive Irish live television rights for Republic of Ireland home
internationals. Seeking to leverage subscription growth, Sky had offered €7.5
million for the games, far exceeding RTÉ's €1.6 million offer. Public reaction
was vociferous: viewers indifferent as to what channel they watched
Champions League games on were furious at the prospect of having to pay

an overseas broadcaster to watch the national team. The government expressed disappointment but resignedly described the deal as, in the words of the minister for sport, John O'Donoghue, 'done and dusted'. However, in the same month, the European Commission emerged as a white knight, expressing the view that such events could be protected retrospectively. Although it was unclear whether this advice was legally sound, the Irish government felt obliged to respond and belatedly – in October 2002 – published a list of protected events including European Championships and World Cup qualifying matches, the all-Ireland hurling and football finals, the Irish Derby and the Irish Grand National. To ensure that the list applied retrospectively (and thus applied to the FAI/Sky deal), new legislation was introduced in April 2003. The Broadcasting (Major Events Television Coverage) (Amendment) Act 2003 stated that it applied to 'a designated event which is designated *before or after* the passing of this Act [italics added]'.

To avoid the possibility of an uncomfortable confrontation with Sky, a subsidiary of News International that enjoyed some political influence both within and without Ireland, Article 4 of the amending legislation shifted the focus of the exercise of the state's authority from the broadcasting organizations to the events organizers. Having finally secured a restoration of the national team's competitive games, in February 2005 RTÉ signed a €5 million deal with the FAI securing live broadcast rights of ROI home international matches up to end of the 2010 World Cup qualifying campaign.

However, if viewer access to certain events was protected, this did not guarantee they remained the exclusive property of RTÉ. As the decade progressed, more and more events went to competing Irish stations. Key among these was Setanta Sports, which had developed from offering Irish sporting content to diasporic communities outside Ireland in the early 1990s to becoming an element of the NTL cable package from August 2004, where it screened Scottish Premier League and UEFA Cup matches. In May 2006, having secured capital through private-equity groups such as Goldman Sachs, Benchmark Capital and Doughty Hanson, Setanta spent £392 million on two of six English Premiership packages – breaking the monopoly Sky had enjoyed since 1992. This was a high-stakes and a high-risk strategy: though Sky had demonstrated the power of the Premiership to build subscriber numbers, Setanta would have to increase their UK subscribers from 150,000 to 750,000 to fund the rights acquisition. These stakes were raised even higher in October 2006 when the channel spent €150 million to outbid Sky for seven years of the UK and Ireland rights to US PGA tournaments.

In Ireland, Setanta beat RTÉ to Formula 1 rights in November 2004 and outbid it again in May 2006 to acquire Irish live rights for midweek Premiership games. In January 2006, after two decades of screening successive Rugby World Cups, RTÉ found itself outbid by Setanta for the 2007 tournament. That the Ireland World Cup games were protected for free-to-air broadcast under the 2003 act, and that Setanta did not reach the 98 per cent population threshold required for listed events, raised the prospect that RTÉ could at least acquire the national team's games. However, they lost again – this time to TV3 – and in a perverse grand slam, in October 2006 the radio rights went to Newstalk 106.

Even GAA rights were not sacrosanct. Throughout the twentieth century, the development of the state-owned broadcaster and of Gaelic games in Ireland had seemed inextricably linked. However, by autumn 2007, an increasingly rights-savvy GAA divided what had previously been a single block of broadcast rights into twenty-five separate packages, split by medium, championship and code. Although key events such as the later stages of the Sam McGuire and Liam McCarthy cups remained on RTÉ, others were shared with TV3 and Setanta.

Some events with a national dimension simply disappeared from free-to-air television. Sky acquired the rights to rugby's Heineken Cup in 2006, precisely the point when Irish provinces started to dominate the tournament (winning in five of the seven years between 2005 and 2012). Similarly, although in autumn 2005 the minister for communications, Noel Dempsey, suggested that the Ryder Cup – due to be held in Ireland the next summer – might be listed as a protected event, Sky made it very clear that it would resist anything that interfered with its exclusive access, and Dempsey did not pursue the matter.

The impact of such intense competition was mitigated by the fact that from 2004, driven by a booming economy, overall broadcasting revenues expanded dramatically, and, to an extent, the rising tide lifted all boats. Figures from the Institute of Advertising Practitioners in Ireland (IAPI) suggest that radio ad-spend increased by 61 per cent between 2003 and 2007, from €87 million to €140 million. Television ad-spend in the same period grew even faster – by 92 per cent – from €195 million to €374 million. By April 2005, ad agencies were reporting that even with fourteen channels to choose from, demand for TV advertising space was outstripping supply. Inevitably this drove up rates. A December 2005 survey, noting that Ireland had witnessed the highest increase in television advertising costs in Western

Europe in 2005, concluded that it had become the '10th most expensive country in the world for buying TV space … 80 per cent above the global average' (*IT*, 29 Nov. 2005). This was reflected in financial results. A 16 per cent year-on-year increase in television ad revenues saw UTV Media profits increase by £3 million to £13.9 million in 2004. UTV's expanding Southern Irish radio division alone earned €8.5 million in 2006. For its part, Today FM in 2004 recorded 27 per cent year-on-year increase in their pre-tax profits of €4.2 million. Although not all private stations enjoyed such success – Spin and Newstalk continued to find it difficult to crack the crowded Dublin market – the independent sector as a whole generally prospered after 2004.

As for RTÉ, May 2004 saw the announcement of a surplus – its first since 1998 – of €2 million for 2003. This was entirely driven by the December 2002 licence-fee increase, which, even after Sound and Vision fund top-slicing, contributed an additional €43 million to the station's revenues. Thereafter, although the small annual licence-fee increases and the property-boom driven increase in television households saw RTÉ's licence-fee revenues increase by a third between 2003 and 2007, this was far outstripped by the pace at which the station's commercial revenues expanded: from €155 million in 2003 to €245 million in 2007. Surpluses at the station surged from €2 million in 2002 to €26 million by 2007.

In October 2004, responding to complaints from private players about RTÉ's advertising practices, Noel Dempsey appointed a team of consultants to review the TV licence fee and its impact on private-sector broadcasters and the advertising market. The resulting report, published in December 2004, echoed the conclusions of an earlier competition authority investigation. The consultants found that RTÉ had not used licence-fee revenue to suppress advertising rates (indeed, they asked why anyone would have thought this to be in RTÉ's interests). Furthermore, they found no evidence that RTÉ was acting in an anti-competitive manner in relation to purchasing programming rights. Indeed, the report emphasized that TV3, which serially complained of RTÉ's practices, had through its ownership structure, access to a range of UK programming (ITV programmes) effectively unavailable to RTÉ.

However, the report was also notable for its inclusion of a key underlying assumption: 'the essence of the remit laid upon RTÉ is that, if RTÉ were the only television or radio channel that Irish viewers could receive, it should be complete and satisfying for the whole population, including minority interests as a whole'. Whether that assumption held, given that RTÉ was obviously no longer the only service available, was not addressed. The consultants' work was facilitated by the manner in which RTÉ had reorganized the presentation of its accounts. The first appearance of the new integrated business division reporting structure in June 2004 (referring to the 2003 financial year) revealed the news and current-affairs-heavy Radio 1 and the performing groups constituted the major drains on resources. Their resource-intensive natures made this unsurprising. Network 2's stand-alone deficit of €10.4 million was perhaps more unexpected. (This may have influenced the August 2004 decision to restore the channel's original name, RTÉ 2; apparently some audiences didn't recognize Network 2 as an element of RTÉ.) RTÉ1 Television emerged as the station's cash cow, earning €63 million in ad revenue, but it was another surplus-producing division – 2FM which contributed a net €3.7 million to RTÉ's coffers – that revived a question from the Ray Burke era. If 2FM was commercially viable, why was it part of RTÉ at all? Such questions were informed by a public-service-as-corrective-to-market-failure frame. RTÉ responded that 2FM was simply the musical extension of the station's public-service remit, addressing musical niches others might ignore, but the 2003 annual report also tacitly acknowledged that 2FM's surplus acted as a cross-subsidy for losses elsewhere in the organization.

The general buoyancy of the advertising market encouraged the BCI to offer new licences. No fewer than sixty-one parties expressed an interest when the BCI invited applications for up to four Dublin radio services in July 2003, leading Today FM to question 'the pace' at which licences were being issued. Nonetheless, by autumn 2006, three more mainstream stations were on air – Phantom FM in Dublin and two regional stations serving the south-west (the Denis O'Brien-backed Spin SW) and the north-west. Two Christian stations also went on air in 2006 and 2007: Life FM in Cork and the quasi-national Spirit Radio.

A potential lifeline for Newstalk appeared in spring 2005 when the BCI advertised for expressions of interest in a 'quasi-national' service. Denis O'Brien's share in the station had increased to 57 per cent by 2004 as the station accumulated losses. However, since the costs of running a quasi-national operation did not significantly scale up from a Dublin service, Newstalk bid for and duly won the licence. With access to 92 per cent of the population, it effectively joined RTÉ and Today FM as a third national station from September 2006.

There were also new television stations. In July 2004 the BCI had called for expressions of interest in new television services (including community TV).

In January 2005 City Channel launched on the NTL Digital platform. Initially limited to 90,000 homes in Dublin from February 2006, it extended to Galway and Waterford via the NTL platform. Relying on low-cost but largely locally produced magazine formatted shows, its schedule addressed niche audiences – the gay community and Polish immigrants, for example – not noticeably served elsewhere. Notwithstanding losses of €350,000 in 2006, City attracted the attention of the transnational cable operator Liberty Global, which acquired a 35 per cent stake in the station for €2 million in August 2007. The investment fuelled City's ambitions: a promised expansion to Cork and Limerick was augmented by plans to expand into Liberty Global's Eastern European markets.

The more significant addition to the local television market was the May 2005 announcement of the launch of Channel 6. Despite low start-up capital of €10 million (from a combination of the Barry family – owners of the tea brand – the Gowan Motor Group and two venture-capital funds) the promoters initially asserted that Irish content, including independently produced material, would be a key component of their schedule. Asserting that TV3's reliance on ITV simulcasts had seen that channel's audience profile age, the new channel would target youth. However, it was Channel 6's route to licensing that distinguished it more than anything else. The promoters initially secured a licence from the UK regulator Ofcom, proposing to broadcast into the Republic from Belfast while potentially evading BCI content codes – precisely the concern Dermot Ahern had raised with the European Commission in 2003. Although the promoters stressed that in practice they would respect BCI regulations on advertising aimed at children and for alcohol, they acknowledged that the UK licence would allow them to evade other potentially onerous BCI service obligations.

Channel 6 would secure a BCI licence before going on air in 2006, but – remarkably – although continuing to assert it would source one-third of its programming from Ireland, its schedules were devoid of news and current-affairs content, a lacuna that the BCI appeared willing to tolerate, perhaps reasoning that the Ofcom licence meant Channel 6 could stay on air regardless of BCI obligations. At the March 2006 channel launch, the only shows mentioned were US acquisitions, and founder Michael Murphy proclaimed that the schedule would be 'no news, no sport, no children's programming, just pure entertainment' (*IT,* 8 Mar. 2006).

As a consequence, Channel 6 was indistinguishable from the plethora of similar channels available to digital viewers, and having initially targeted a 3 per cent audience-share, the opening weeks saw it average just 1.1 per cent. While the rest of the industry thrived, Channel 6's management was reduced to contemplating screening soft-core porn, and in October 2007 the channel closed its in-house advertising-sales team, switching to that of TV3, foreshadowing a much closer relationship.

By this stage, TV3 was in new hands. In January 2006, a debt-heavy Canwest, mindful that a 20 per cent increase in Irish television ad-spend in 2005 made the station an attractive purchase, announced that it was up for sale. As co-shareholder, ITV (which had emerged from Granada's 2003 merger with Carlton) had first refusal, but proved uninterested. UTV and Setanta took a keen interest, but it was a London-based venture-capital fund, Doughty Hanson, which secured not just Canwest's 45 per cent stake in May 2006, but subsequently that of ITV and the original investors (Paul McGuinness, Ossie Kilkenny and James Morris), with a €265 million bid in August 2006. The threatened loss of some key ITV shows was averted by a long-term deal signed in December 2006 to retain *Coronation Street* and others. Surprising those who assumed that Doughty Hanson's interest in TV3 would be short-lived, in February 2007, the new chief executive, David McRedmond, announced that TV3 would increase its home-grown programming under the gaze of Ben Frow (ex-Channel 4) as head of programming. The first instance of this, a nightly entertainment news show called *Xpose*, was followed by a surprisingly lively addition to current-affairs output on Irish television. After RTÉ axed his late-night Radio 1 show, veteran journalist Vincent Browne brought his nightly take on current affairs to TV3 from August 2007. Browne continued to dog at evasive politicians on the airwaves, and, in a deliberately expansive guest strategy, brought hitherto unheard and unseen voices and perspectives to Irish screens in a manner so successful (relative to the late-night slot) as to lead RTÉ to cancel plans for a similar show.

In a bid to extend its audience reach, February 2007 saw Channel 6 – hitherto exclusively available on NTL – negotiate carriage on the Sky platform. Even so, its ratings remained poor, and the station must have looked longingly at the 180,000-plus homes that remained beyond its reach because of the ongoing disarray in the government's digital plans. Though the 2002 forum had recommended that DTT be revisited, a January 2004 report commissioned by the department of communications concluded that a lack of private-sector interest meant the state would have pay anything between €40 million and €100 million to build DTT infrastructure. Under pressure from the EU to complete the migration from analogue to digital broadcasting, the government supported a DTT pilot project developed by BT, NEC and RTÉ, which launched in Dublin and Louth in 2006. A year later, in April 2007, the Broadcasting (Amendment) Act created a legal basis for DTT, which essentially left its development to RTÉ (as network operator), the BCI (to licence a commercial multiplex operator) and ComReg (to allocate spectrum).

In the interim, the development of the digital-television market was left to commercial players. By early 2004, even without DTT, Ireland had the second-highest digital-television penetration (28 per cent) in Europe, courtesy in part of NTL and Chorus, but mainly Sky, which grew from 355,000 digital subscribers in May 2005 to 465,000 by December 2006.

The potential for cable to mount a meaningful challenge to Sky relied on new investment, which began to arrive in 2005 after a complicated series of mergers and acquisitions. In January 2004, Independent News and Media offloaded Chorus onto US cable giant Liberty Media, which in turn sold it on to Europe's largest cable firm, UGC, the following December. A month later, in January 2005, UGC moved to acquire NTL's Irish operation, but as it did so, Liberty Media and UGC announced merger plans creating a new entity, Liberty Global, and it was this entity that acquired NTL Ireland for €330 million. The acquisition was not straightforward from a regulatory perspective, however, and not only because Liberty's prior ownership of Chorus (via UGC) would create a cable monopoly in Ireland. That Liberty owned 18 per cent of News Corporation, which in turn owned 39 per cent of Sky, created a potential satellite-and-cable-television distribution monopoly, which delayed competition authority approval of the deal. Once approved though, Liberty rebranded Chorus/NTL as UPC, beginning the investment push to finally complete the upgrade of the cable network.

The consolidation of ownership within the television-distribution market reflected a similar pattern evident in the radio-station market since 2000. This was facilitated by changes to the IRTC/BCI's ownership and control policy in 2001, prompted in part by a November 2000 UTV bid for County Media, which operated three commercial stations under the 96FM and 103FM brands in north and west Cork. While UTV sought to acquire 100 per cent of County Media, the IRTC's existing ownership policy limited the share any single entity could hold in a radio station to 27 per cent. UTV pointed out that such limitations were not applied outside the radio market. By 2000 the IRTC had already permitted both Canwest and Granada to establish holdings of 45 per cent in TV3. There were also other apparent ownership-policy inconsistencies: that Radio Ireland (owner of Today FM) accounted for 40 per cent of the shares in consortia applying for Cork and Dublin licences in 2000 and 2001 had not prevented the IRTC from short-listing those consortia for the licences.

Nonetheless, the IRTC initially refused to approve the County Media deal on the grounds of Section 6 (2) of the 1988 Radio and Television Act, referring to the character of the applicant. The IRTC stated that it favoured strong local shareholdings in radio services, and that the UTV bid 'represented a fundamental shift in that policy' (*IT*, 30 Jan. 2001). A month later, however, the IRTC permitted UTV to acquire a smaller – 60 per cent – stake in County Media, on the condition that the remaining 40 per cent be retained by the existing shareholders. Regardless, the decision publicly signalled that the 27 per cent limit was no longer inviolate, and that controlling stakes were potentially available.

In a tacit acknowledgement that practice had superseded the rules, in May 2001 the IRTC began a consultation on media ownership. It received thirty-three submissions from cable operators, government departments and a large number of private broadcasters. Key among them was that of the competition authority, which argued that cross-media ownership restrictions should be relaxed because their 'too rigorous application ... could have the unfortunate effect of stifling the flow of capital to broadcasters', adding that waivers to the rules should be granted where 'public interest benefits resulting from cross ownership and the resultant capital flows outweigh the threat to diversity and localism' (*IT*, 1 Sept. 2001). The authority was echoing a liberalization discourse that was increasingly prominent across Europe. Writing in 2000, Gillian Doyle noted that 'restrictions over the size of media firms ... have been portrayed as detrimental to the economic interests of the industry and this view appears to have fuelled an international trend towards deregulation of broadcasting and press ownership' (1–2).

By its nature, the competition authority was unlikely to ever prioritize diversity of content and localism as core considerations, over an emphasis on economic efficiency. The submissions from UTV and Scottish Radio Holdings (which owned 24 per cent of Today FM) also sought a liberalization of controls, SRH arguing that existing restrictions worked against 'the interests of the listener, employee and the growth of a dynamic independent broadcasting sector' (*IT*, 1 Sept. 2001). In sum there was a strong chorus of voices seeking liberalization, with only a handful of submissions raising doubts about such a direction. Overall, the BCI noted 'strong support' for the regulator to focus on content rather than ownership as a means of ensuring that broadcasters retained a local ethos.

Although the BCI did not belabour the point, this implied close monitoring of the content. Whether such a labour-intensive task lay within the capacity of the BCI was not discussed. The BCI summary also noted that most respondents considered the focus on the existing ownership limits as too narrow. The precise percentage shareholding in a media firm mattered less than the influence exerted by a shareholder. If, for example, SRH's minority holding already gave them a dominant voice in discussions on the future direction of Today FM, what did it matter whether they owned 24 per cent or 100 per cent?

In October 2001, the BCI published a new code on ownership and control. The BCI explicitly stated its intention to adopt 'a lighter touch' to regulation, 'while adhering to its statutory obligations of ensuring pluralism and diversity'. More specifically, the new code abandoned any percentage limits on radio shareholdings within a single broadcaster. These were replaced by limits applying to the radio industry as a whole. In effect, the new rules stated that no one company should own more than a quarter of all radio stations licensed by the BCI.

The decision was not without its detractors, who pointed out that the IRTC had already overseen a relaxation of Today FM's initial content obligations (away from a news and current-affairs focus in its original incarnation as Radio Ireland to a lighter, magazine format following the intervention of Chris Evan's Ginger Productions), while TV3's reliance on imported content substantially diverged from the schedule envisaged in its original 1989 submission. Nonetheless, by permitting a single entity to own up to 25 per cent of the entire radio market, the context for media ownership in Ireland was immediately transformed.

UTV and SRH immediately moved to acquire those shares in County Media and Today FM they did not already hold and, over the next few years, built larger radio groups. In April 2002, UTV spent €15.74 million on Limerick station Live 95FM, and in December 2002, it spent €14 million on Dublin-based Lite FM (later rebranded as Q102). Two years later, in October 2004, it added LMFM, the largest local station outside Dublin and Cork, at a cost of €9.5 million. The five-station Republic of Ireland group (together with the Belfast-based U105, launched in 2005) made a significant contribution to the health of UTV's overall finances. In 2003, over 12 months, the

UTV Media group earned pre-tax profits of £9.5 million. By September 2006, it was reporting pre-tax profits of £8.5 million for just *six* months. Nor were UTV's ambitions limited to the island of Ireland: in June 2005 it acquired the UK-based Wireless Group (with its flagship national Talksport stations) for £97 million.

SRH, which in addition to Today FM had built up significant press holdings – the *Leitrim Observer* (1998), the Kilkenny People group (2000), *Ireland on Sunday* (2000) and the *Longford Leader* (2002) – also continued its radio-market expansion. An initial approach to the loss-making Country 106FM in Dublin in September 2002 came to nothing, but a €26 million bid for FM104 in October 2003 was quickly approved by the BCI. In June 2005, it also acquired Highland Radio for €7 million.

Such empire building was not limited to overseas groups. A week after winning the second Cork radio licence with Red Hot FM in February 2001, Thomas Crosbie Holdings had purchased the Leinster-based Provincial Papers Group, which included the *Nationalist and Leinster Times*, the Kildare *Nationalist*, the Laois *Nationalist* and a Carlow paper. A controlling (75 per cent) share in WLR and in Beat FM was added to these in April 2007. In February 2005, Radio Kerry announced a move to acquire Shannonside/Northern Sound from the Midlands Radio Group for approximately €6 million; a subsequent (December 2007) move to acquire Clare FM for €7.2 million did not proceed. Even the *Connaught Tribune* entered the fray in 2006, buying out the 73 per cent of Galway Bay FM it didn't already own in a transaction valuing the station at €20 million.

Some of the valuations placed on the stations in these deals perplexed observers. Today FM, for example, had made a profit of just €400,000 in the year preceding SRH's decision to offer €26 million for the station. Clearly the new owners anticipated – correctly in the case of Today FM, as it turned out – that there were greater profits to be made. In November 2003, Dermot Ahern's department of communications commissioned OX consultants to examine whether the state could benefit from these purchase prices on the grounds that such stations were only possible because the state had granted them access to publicly owned frequency spectrum. Unsurprisingly, there was significant push-back from the private sector, which emphasized that, far from being a licence to print money, the average net profit of Irish independent radio broadcasters was just €300,000, and some new entrants to the Dublin market in particular were struggling to survive.

The pace at which radio ownership was becoming concentrated raised at least temporary concerns at a regulatory level. In December 2004, the BCI imposed a temporary freeze on any investor in commercial radio stations holding more than 17.9 per cent of the market – exactly the level UTV had just reached with LMFM, although Denis O'Brien's Communicorp was not far behind at 14 per cent. Simultaneously, the competition authority took the unusual – for them – step of examining SRH's acquisition of Highland Radio at length, on the grounds that its prior ownership of Today FM and FM104 conferred potentially significant market power on SRH. Although the authority approved the deal in August 2005, it did so subject to the insistence that Highland end its participation in Independent Radio Sales, a sales-and-marketing operation shared with sixteen other commercial radio stations. For its part, having announced a review of ownership and control policy, the BCI was content to lift the 17.9 per cent cap in November 2005 in anticipation of the new policy's publication.

In the event, this was superseded by the largest broadcast acquisition yet. In June 2005, Scottish Radio Holdings' UK and Irish holdings were acquired for £391 million by Emap, one of the UK's largest media groups. Emap immediately sold SRH's print holdings to UK regional publisher Johnston Press, while retaining the broadcast operations. Two years later, in February 2007, Emap placed its Irish radio stations – Today FM, FM104 and Highland Radio – on the market for €200 million. Though this was a hefty price tag, all three stations were highly profitable. In January 2007, FM104, rumoured to be the single most profitable local broadcaster in Ireland, filed €1.5 million in profits for the six months up to March 2006. Today FM and Highland reported pre-tax profits of €2.9 million and €500,000 for the same period. Unsurprisingly, the sale attracted substantial interest: fifteen companies sought the prospectus, which emphasized the favourable economic conditions and the 9 per cent compound annual growth rate in radio advertising between 2002 and 2007. Six companies – including UTV, TV3, Communicorp, New Wave Media, the Irish Times and a private equity group – were shortlisted. In July 2007, it was announced that Denis O'Brien's Communicorp had prevailed, in a deal worth €200 million.

This constituted by far the largest Irish acquisition to date for Communicorp. Until this point, it had been involved in its existing stations – 98FM, Newstalk, Spin and Spin South-West – from launch, building them from the ground up. Prior to the deal Communicorp held five of the thirty-three radio licences, or 15 per cent of the total. The addition of the Emap stations brought this figure to 24 per cent, just under the 25 per cent limit enshrined in the BCI's ownership and control policy. However, concerns were raised by the nature of the stations involved. The deal raised the prospect of a single company owning both national commercial broadcasters *and* dominating the Dublin market, where the combined market share of FM104 and Communicorp's existing stations was over 53 per cent. Inevitably, the acquisition attracted close attention from the competition authority (which in November 2007 commenced a Phase 2 investigation, a mechanism rarely used in previous Irish media mergers) and the Broadcasting Commission of Ireland. The deal was further complicated by Denis O'Brien's acquisition of a growing stake in Independent News and Media, which had reached 8.6 per cent by July 2007. This continued to build even as the competition authority and the BCI contemplated the Emap acquisition, reaching 11 per cent by November 2007.

In the event, it was a BCI intervention that proved decisive. In October 2007, it announced that it would not approve the transfer of another Dublin licence to Communicorp. This tied the hands of the competition authority. When it granted its approval, it had to be subject to the condition that Communicorp immediately sell off FM104. Another competition was held for FM104, in which UTV prevailed with a reported December 2007 offer of €52 million.

Over the course of the years between 2000 and 2007 the Irish broadcasting market had changed beyond recognition. The multiplicity of owners present at the turn of the century had been significantly reduced in both transmission infrastructure and radio stations. The creation of mini media empires at increasingly eye-watering costs – UTV's 2007 offer for FM104 was seventeen times the latter's last reported profits – pointed to substantial faith in the capacity of the market to not merely maintain existing revenues but to significantly expand them. It remained to be seen whether such faith could be justified.

10 Into the abyss: print media, 2007–16

The print media in Ireland, in the period between the beginning of the economic collapse that affected Ireland in 2007–8 and 2016, went through as dramatic a transformation as many other aspects of society and the economy in that period. It was a period that was also marked by developments in the relationship between government and media, by regulatory changes, by some extraordinary defamation cases and by a deepening academic interest in the study of Irish media in all its aspects.

Some circulation declines were precipitous. The *Irish Sun's* sales fell by almost 48 per cent, and other daily tabloids fared little better. These declines were evidently multi-factorial, but the simultaneous decline – virtually a collapse – in the building industry led some to suppose that there was a relationship between the fall in tabloid circulation and the disappearance of the building sites to which working-class male newspaper purchasers would traditionally have brought their daily newspapers.

The overall circulation changes between 2000 and 2015 were dramatic (all figures relate to the January-to-June period). The two strongest daily papers, the *Irish Times* and the *Irish Independent,* were down by 34 per cent, and the *Sunday World* by 42 per cent. Effectively, the only outlier was the *Farmers Journal* which, benefiting from its unique status as a niche paper, fell by only 0.5 per cent.

There are other elements of these statistics that are also worth noting, but for different reasons. The disappearance of the *News of the World* from the Irish and UK markets in 2011 was of course related to the phone-hacking scandal that led to the Leveson inquiry into this and other newspaper practices, and was not as a result of economic changes. In Ireland (as in the UK) the disappearance of the *News of the World* led to the introduction of a Sunday edition, in both the UK and Ireland, of the *Sun* newspaper, also part of the Rupert Murdoch empire. This was only partly successful in filling the gap left by the *News of the World,* which had, prior to its closure in 2011, been selling 115,000 copies weekly in Ireland. Circulation figures for the *Irish Sun on Sunday* were not compiled until July–December 2012, when it was selling 65,000 copies a week, but it was demonstrating a degree of resilience not obvious elsewhere: in the 2012–15 period, its circulation fell by only 14 per cent.

The *Independent* group continued to support the ailing *Sunday Tribune* until its unsustainability became obvious in early 2011. The newspaper ceased publication and a receiver was called in. Immediately afterwards, the *Irish Mail on Sunday* published, on 6 February 2011, 25,000 copies of a 'wraparound' or facsimile version of the *Sunday Tribune*'s front page, enclosing its own publi-cation. The Independent group contemplated suing the Mail group, but did not do so; however, the receiver took the Mail group to court, and his case for 'passing off' was settled in July of the same year in the commercial court for an undisclosed amount.

The Daily Mail group's involvement in the Irish mid-market newspaper wars was becoming intense in this period, buttressed by an apparently bottomless supply of funds. In the decade after 2006, the cumulative losses of the two Irish Mail titles amounted to over €50 million. On the other hand, it may well have been the case that investment in circulation in Ireland had a below-the-line benefit for the group, in that it may have been more cost-effective than investing in circulation growth in the UK, and of course also because it opened up a potentially lucrative new advertising market in Ireland. The *Irish Daily Mail* launched in 2008, and by 2015 its circulation had fallen by a relatively modest 15.8 per cent (although this is not strictly speaking compa-rable with the circulation losses of other papers since 2000, as it was for a shorter period).

At Thomas Crosbie Holdings (owner of the *Irish Examiner*), the post-1990s print and broadcast acquisitions spree began to look ill-advised as the recession struck. A substantial restructuring of the company followed, with the original Thomas Crosbie Holdings (run by five generations of the Crosbie family) going into receivership in March 2013 and being succeeded by Landmark Media Investments, owned by the same family. Although its financial expansion was paralleled by a period when it trumpeted its claims as a national paper – and its editorial independence and breadth of news coverage to some degree warranted this description – in more recent years its policy has reflected the importance of its circulation and business heartland in Munster. There, its chief rivals in its mid-market field have been the *Irish Independent* and, to a lesser extent, the *Daily Mail*.

Local newspapers were not exempt from the challenges of calibrating their cost bases to the capacity of the market to deliver revenues. In one extended case history (Cawley, 2016) the author pointed out that Johnston Press had paid more than €200 million in 2005 to acquire fourteen local or regional Irish titles, and sold them later for just €8.5 million.

In 2008, Local Ireland was founded as an organization amalgamating the former Provincial Newspapers Association of Ireland (formed 1919) and the

Irish Master Printers' Association. The new organization incorporated forty-six weekly paid-for local newspapers of record, with a combined circulation of around half a million copies and a weekly readership estimated at 1.4 million. Some national news stories – i.e. the child-abuse scandal in the Co. Wexford diocese of Ferns – were broken by the local newspapers in their areas. While their circulations have been in decline, those declines have been less marked than those of the national print titles, and, while the challenge to monetize online content remains, member titles experienced double-digit annual growth in their unpaid online readership. Cawley concluded:

> Speaking of local journalism's current resource challenges is not to pine for a golden age now passed. Irish local newspapers have always operated on limited means and imperfectly performed watchdog and informed-citizenry roles. But their significance has been in offering institutional frameworks and support for journalists systematically to gather and report local news … The alarm now is that editorial resourcing is so tight it is choking local journalism's democratic contribution. (Cawley, 2016: 86)

The newspapers were generally experiencing a perfect storm but, as was not effectively realized until later, some of the winds that were buffeting them were of their own making, as they migrated their expensively generated content to the internet, where it instantly became available to consumers for nothing. Initially, however, the experiment of going online appeared popular, if not profitable, particularly among the Irish diaspora. One 2001 study demonstrated that two-thirds of the contacts made at that time with the *Irish Times* portal, Ireland.com, originated from abroad.

The *Irish Times* had been free online from 1999, but from 2002 charges and a registration fee were introduced for most of its content. The following years saw considerable experimentation with attempts to monetize its content. In 2008, the newspaper was free online, but charged for access to its archives. In 2015, it followed the practice of many other newspapers internationally by introducing a 'leaky' paywall: a limited amount of content was available without charge, but a subscription to the online edition was required for full access. The period was also marked by a number of commercial decisions that, in retrospect, could be seen as problematic. These included the purchase in 2006 – just before the property crash – of Myhome.ie, a property website, for more than €40 million. This was evidently intended as a way to profit from the migration of property advertising to the internet, but two years later the parent Irish Times company was registering a loss of almost €38 million, and there was little prospect that a cycle of retrenchment and job cuts at the paper would end any time soon.

The move online by traditional print media was not rapid, other than in the sense that publications' hard copy versions were readily transferable onto the internet. Although the Dublin titles of the *Irish Independent* group first went online in 1997, the *Independent.ie* domain was not launched until 2001. Further major developments took place in 2007, when the newspapers' Facebook presence was launched. The papers' online presence moved to the Escenic EMS platform in 2006–7; its app was launched in 2008; and it launched on Twitter in November 2009. Site redesign took place in 2013, and there was a further redevelopment throughout 2014–16. A number of key metrics are not available because they are commercially sensitive.

Adaptation to the internet by individual journalists and by news-gathering print organizations was hesitant and partial. One early study noted that:

> while journalists showed some appreciation for the potential of the internet to foster content diversity, they also held the view that existing, mainstream media structures would remain intact. Similarly, while recognizing advantages such as speed and archivicity, most favoured continuity over a radical break with previous content formats. (O'Sullivan, 2005: 66)

Tension between the political and media establishments is, of course, a constant. It was nowhere more evident than in the lengthy controversy between 2006 and 2009 involving a major clash between the right of a newspaper to publish material in the public interest and the right of a judicial tribunal to protect confidentiality in the course of its own proceedings.

In the same time period, a number of high-profile libel cases led the print media in particular to intensify their pressure on the government in relation to hoped-for changes in the defamation law, which had effectively remained unchanged in Ireland for almost half a century, and which was largely based on English law from the same era.

These cases included a 2008 award of €900,000 to someone described in a newspaper as a 'drug lord', but who had no convictions for drug offences; this decision was overturned on appeal. At the time of the original decision, the industry body National Newspapers of Ireland, made a submission to the joint committee on the constitution arguing that awards of this magnitude cultivated a chilling impact on freedom of expression due to the availability of large damages, combined with the unpredictability of the system that leads to their assessment. An appeal process in relation to this award was, at the time of publication, ongoing.

This was also the case with the jury award of €1.87 million to a public affairs consultant, Ms Monica Leech, for statements published about her in a number of newspapers of the Irish Independent group in June 2009. This was considered disproportionately high by the supreme court, which reduced it to €1.25 million in 2014, but the newspaper group has appealed to the European courts against this verdict, and, at the time of writing, this case has not yet been resolved. This illustrates the extraordinary length of time (and unquantifiable legal fees) that can be consumed by defamation cases. It was scant consolation for the newspaper industry that the largest single defamation award in this period was not against a newspaper, but for €10 million against a company for material published in a press release in November 2010.

Politicians in Ireland have not been slow to take actions against newspapers for defamation, but many of these cases are settled privately, so it is difficult to assess their significance. The most high-profile case involving a politician – in this case deceased – was taken during this period by a female Russian employee of Liam Lawlor, the TD and businessman, against the *Sunday Independent* and the *Sunday Tribune* for references to her following a car crash in Moscow in which her employer, who was also a passenger, had died. The initial basis for this was a dispatch from a freelance journalist in Brussels, who contributed a similar report to a London newspaper, which also settled a defamation action. This case was brought in 2007, shortly after the accident, and was settled for an undisclosed sum two years later, despite the fact that the woman concerned had not been named or positively identified. An internal report on what had happened was carried out by the Independent group, but not published (Cullen, 2007). One commentator suggested that the reason the newspapers had 'coughed up' was 'not fear that the case would be successful and that Kushnir would be awarded millions in damages, but that the editorial standards that prevailed, notably in the *Sunday Independent*, *Sunday Tribune*, and other newspapers, would be exposed' (Browne, 2007). In one of the few cases in which a journalist was vindicated, Sam Smyth in 2012 successfully defeated an application by the politician Michael Lowry for a summary decision against him for comments he had made about that politician's business dealings.

Pressure from these and other cases resulted in the government finally agreeing to introduce a new defamation statute, but, while some of its provisions were welcomed, it was still seen by some academic lawyers as inadequate. Another problem was that, towards the end of the government that introduced it, the cabinet decided to make passage of this legislation conditional on the passage of proposed new legislation on invasion of privacy (*IT*, 11 May 2009), drafted originally in 2006. This was evidently seen by some members of the cabinet as a necessary counterbalance to the greater protection afforded to media in the defamation statute, and underpinned the general wariness among professional politicians about offering any kind of an olive branch to perennially critical media without exacting a suitable price in return. Media interests were markedly hostile to this development.

Another (briefly controversial) aspect of the legislation, which was not eventually passed until 2009, was that two of its sections – 36 and 37 – made blasphemy a criminal offence. This was an apparent solution to an old problem caused by the failure of a number of private prosecutions for blasphemy because, while blasphemy had been proscribed in the 1937 constitution, it had never been defined in law. Faced with the alternative of successfully promoting a national referendum to delete the blasphemy clause from the constitution, and defining blasphemy by statute, this decision by the government represented the softer option – particularly as the length of the list of conditions to be satisfied in any successful prosecution rendered the likelihood of any such prosecution satisfyingly remote. National Newspapers of Ireland argued unsuccessfully in 2009 that 'the matter be best dealt with by Constitutional Referendum either to retain or amend the provisions of Article 40 (6) (1) (i)' (Cullen, 2009: 6).

Debate about the defamation and privacy proposals continued to some degree in tandem, and revealed some sharp disagreements, which rapidly diluted the qualified welcome the industry had given to the defamation proposals. A lawyer for the *Irish Times* rejected the government's contention that there was a need for privacy legislation, and argued that the proposed press council and press ombudsman, as provided for in the defamation action, was sufficient to deal with 'professional thuggery let loose and conditioned by a newspaper proprietor' (*IT*, 2008). The interpretation and construction of the proposed legislation, he commented, would be 'a lawyer's delight'. Although the defamation legislation was not passed before the 2007 general election, its connection with the proposed privacy law had been severed, probably in part because of the opposition by media interests.

Although there were differing opinions about the utility of the new, or at least newly formulated, defences available to publications following the passage of the 2009 Defamation Act, there was more general welcome for its proposals for a voluntary system of press regulation by a press ombudsman and a press council. These two institutions were set up informally by the press industry in 2007, and formally recognized by the Oireachtas in 2010.

Since the institution of the council and the appointment of the press ombudsman, certain significant trends have become obvious. The most significant is that complaints about the invasion of privacy, while more numerous in raw terms than had been anticipated by the research carried out by Boyle and McGonagle a decade earlier, are not the largest category. The complaints most frequently registered about newspaper or magazine articles are on grounds of accuracy. In the first year of operation, for example, complaints about accuracy featured in some 35 per cent of the adjudications; those relating to privacy accounted for 11 per cent. In 2015, the latest year for which figures are available, the percentages for each of these categories of complaint were 46 per cent and 13.7 per cent respectively (PCI/OPO, 2015: 208).

One major piece of academic research – a comparative study of the press council and the office of the press ombudsman (Brody, 2012: 44) – concluded that:

> in order to provide greater awareness of media accountability systems, media literacy curricula in schools and colleges should include a study of such systems, as this would ensure optimal levels of media accountability in the future. Media education of media professionals, especially in light of ever-changing media technologies, is also important to ensuring higher media standards and effective accountability to the public now and in the future. It is therefore recommended that specific education of media professionals should be promoted.

It added:

> The plethora of sources of information available in the twenty-first century means that the relationship between the public and the news media requires more transparency than ever before. The public is ever more reliant on the media to provide it with accurate and reliable information. In order to achieve a relationship of transparency and trust, newspapers and media organizations must ensure that effective media

accountability mechanisms are put in place and reviewed and revised as necessary in light of developments in technology and other societal forces. (Brody, 2012: 278)

Although the future of the proposed privacy legislation remained uncertain, one high-profile case raised privacy issues at common law. A married woman claimed that her right to privacy had been breached by the publication in the *Irish Daily Mail* and the *Irish Mail on Sunday* of excerpts from secretly taped telephone conversations between her and a priest with whom she had an intimate relationship following her separation from her husband. Although the conversations had been published in 2003, the case did not come to hearing until 2008, when the plaintiff was awarded €90,000. In her judgment, Ms Justice Dunne rejected the newspaper's contention that it was merely exercising its right to freedom of expression, saying that she could not see how anyone could assert that right in order to publish transcripts of private telephone conversations where phone-tapping has been expressly prohibited by the legislature. Perhaps equally significantly, she said that if the paper had chosen to write about the conduct of the priest concerned, some information relating to the plaintiff could have been legitimately brought into the public domain, but in the circumstances the newspaper had gone beyond what would have been permissible (Carolan, 2008; Courts Service, 2008).

It became clear that although the legislation had not been shelved indefinitely, the government had to some extent embarked on a search for a level of consensus. Michael McDowell's successor as minister for justice, Fianna Fáil's Dermot Ahern, argued that the issues of privacy were of wider concern than to the major protagonists, and opened the door to 'a wider, structured discussion and further consultation on the issue to ensure the bill achieves [its] ends' (Coulter, 2009). Seven years later, in 2016, there has been little evidence of consultation, and further progress appears to have been stymied both by the intervening coalition government and by its successor Fine Gael/independents coalition.

The main issue in relation to the defamation legislation was that it defined the defence of 'reasonable publication' – a statutory initiative for which media interests had long been calling – in ways that added to, rather than subtracted from, the difficulties faced by media. One legal authority argued that because the act had set out no fewer than ten factors that could be taken into account in determining whether it was fair and reasonable to publish any statement, it was very unlikely that the defence could ever be successful (O'Dell, 2010). He also argued that the statute was possibly flawed both under the Irish constitution and in European law because it did not take account of the McLibel case determined in the European Court of Human Rights in 2005. The ECHR had decided in that case that the Convention on Human Rights had been infringed because of the radical imbalance between the parties, the absence of legal aid for the defendants, the presumption that the statements complained of were false and the fact that the plaintiffs did not have to prove actual damage to their reputation.

The passage of the Defamation Act in 2009 did not have any immediate effect, as many of the defamation cases held immediately following it had been commenced prior to 2009 and, in some cases, remained in the courts system for a considerable period afterwards. One case heard under the new legislation, however, had unusual aspects in that it was brought against a newspaper – the *Irish Daily Mail* – by the largest single shareholder in a rival newspaper group, Independent News and Media, Denis O'Brien.

Paradoxically, the same individual's critique of newspapers had, in the period immediately preceding this action, been directed against some newspapers in which he had a controlling – though not a majority – shareholding, particularly the *Sunday Independent*. Mr O'Brien had built up his interest over a number of years, at considerable expense, because he acquired his shares largely during a period when they were at a premium, before the economic crash that reduced their value dramatically. He was estimated to have lost on the order of €400 million as the value of his purchased shares declined. During this period, the day-to-day operations of the group were effectively managed by the inheritors of the controlling interest of the former chairman, Sir Anthony O'Reilly, whose son Gavin was the company's chief executive officer. In 2011, Mr O'Brien took the unusual step of contributing to a rival newspaper – the *Irish Times* – an article rejecting allegations that he had been exercising improper influence over the papers in which he was a significant shareholder (O'Brien, 2011). This was followed, almost immediately, by a statement from INM also rejecting these claims (*IT*, 17 Nov. 2011).

O'Brien also took a libel action against the *Daily Mail* in relation to an article that had expressed strong opinions about his financial involvement in the operation to relieve distress in Haiti following the 2010 earthquake there. This resulted in a jury award to him of €160,000, suggesting that the jury believed that the article, while it expressed the writer's honest opinion, was neither based on full facts nor in the public interest.

While the newspapers were engaged, during this period, in a perennial conflict with actual or potential litigants, another controversy, which took some eight years to resolve, highlighted the sometimes difficult relationship between the media and the judicial system and, by extension, the state itself. The context was the establishment and progress of the Mahon tribunal, which had been established almost a decade before, in 1997 (under a different judge), to enquire into payments to politicians in relation to planning matters and which, in 2006, was investigating payments made to the taoiseach, Bertie Ahern. In that year, the *Irish Times* had come into possession of documents relating to Mr Ahern that it published on 21 September. Both the tribunal itself and an individual named in the document threatened legal proceedings, the tribunal relying on an injunction against the publication of such documents that had been granted by the supreme court the previous year. On 25 September the tribunal made an order requiring the reporter concerned, Colm Keena, and the editor of the newspaper, Geraldine Kennedy, to produce all the documents on which the article had been based. Battle was joined.

There was little or no precedent for what happened subsequently. The newspaper's legal advisers told the company that they had no option but to comply with the order of the tribunal. But the editor was in a quandary. For one thing, the newspaper believed that all the tribunal's correspondence contained security codes that would enable the tribunal to identify the source of the documents in the possession of the *Irish Times*. On this score alone, the newspaper believed that it had a duty to protect the identity of its informant, and that the public interest in the publication of the information in the documents could be argued to be sufficient to outweigh the minatory approach of the tribunal. There was also another consideration: in a previous UK case, the *Guardian* newspaper, which had refused to hand over to a court official department of defence documents on which it had based an article, was subjected by the British court to swingeing daily fines until it capitulated.

The newspaper therefore decided, against the advice of its legal representatives, to destroy the documents. It informed the tribunal, and the tribunal then sought the support of the high court to force the journalists concerned to answer questions on the matter.

The newspaper argued trenchantly that publishing information from confidential sources was a public good in that it had brought public attention to matters including medical malpractice, improper influence on the planning process, child sexual abuse, improper payments to politicians, tax evasion and the misuse of public funds. The high court, however, found against the newspaper, arguing that its destruction of the documents was 'anathema to the rule of law and an affront to democratic order. If tolerated it is the surest way to anarchy' (Kennedy, 2014; Courts Service, 2007). The newspaper appealed to the supreme court, which, in July 2009, overturned the high court decision inasmuch as – it held – that court had not struck the correct balance between the journalistic privilege derived from the right of freedom of expression and the public interest inherent in the tribunal's need to trace the source of the leaked documents (Courts Service, 2009a). This was, however, a somewhat pyrrhic victory for the newspaper, in the sense that in a subsequent decision in November 2009, the supreme court also held that the newspaper's behaviour in destroying the documents at the core of the case was such as to deprive them of the normal expectation that their victory would be followed by an award of the very substantial costs involved (Courts Service, 2009a). The newspaper was therefore ordered to pay the tribunal's costs in both the high court and the supreme court, as well as its own costs. The tribunal's costs alone in these actions came to some £380,000 (Courts Service, 2009b).

The newspaper in 2010 appealed this final judgment to the European Court of Human Rights under Article 10 of the European Convention on Human Rights, which guarantees the right to freedom and expression of information. The ECHR advised the parties to reach a friendly settlement, but this proved impossible and, in September 2014, a majority of the seven-judge European court rejected the *Irish Times'* complaint as 'manifestly ill-founded' (ECHR, 2014). Although more recent Irish legislation about the rights of whistleblowers has added a certain level of protection (though not explicitly for media), it is clear that the present situation regarding the protection of journalistic sources is not as clear as the media would prefer it to be.

The first decade and a half of the twenty-first century – the period covered by this and the preceding chapter – were also remarkable for the vigorous interest of media researchers generally in applying classical research methodology not only to the historical print media but to the print media's interaction with, coverage of, and influence on, contemporary events. The year 2001 saw not only the publication of the first edition of this book, but also of Mark O'Brien's history of the *Irish Press* (O'Brien, 2001). There were two books about the history of the *Irish Times* (O'Brien, 2008; Brown, 2015), the latter an independent but authorized history, and one on the *Irish Independent* (O'Brien and Rafter, 2012). Other publications focused on Irish journalism before independence (Rafter, 2011), and periodicals and journalism in the nineteenth and twentieth century (Andrews, 2014; O'Brien and Larkin, 2014).

Among the significant academic research studies in the period covered by this and the preceding chapter was a study of accuracy in Irish newspapers, commissioned by the press council in 2007 (Fox et al., 2009). Based to some extent on earlier American methodologies (interestingly, no comparable studies exist for the UK), it was presented as a pilot whose results were indicative rather than definitive. Resource limitations meant, for instance, that while persons mentioned in these reports (more than 500 news reports from fourteen different newspapers were the basis for the survey) could be asked about errors, it was not possible to present the initial findings to the journalists who had written the reports. The response rate was 21 per cent; only 3.4 per cent of the respondents reported that they had found serious or very serious errors (subjectively assessed). The academic categorization of errors in previous studies has divided them into objective (errors of fact) and subjective (errors of meaning) categories. It was interesting that the findings of this study mirrored very closely those of earlier studies. The authors commented:

> Overall, our findings support those of other studies in what appears, at least at first sight, to be counter-intuitive: the more the reporter tries to add to a story, the more likely quoted sources will be unhappy with the results; the less the reporter tries to analyze and contextualize, the more likely the sources will approve of the printed story. In the age of the 24-hour news cycle, newspaper journalists tend to see their role as increasingly one of analysis and less one of recording and conveying raw data. The results of this study and of others indicate that sources (who are also readers) perceive that errors arise exactly in the aspect of journalists' work that is supposedly adding value to information. (Fox et al., 2009: 6)

A substantial body of research, understandably, focused on the role played by print and other media both in relation to the Northern Ireland conflict (which moved into a different phase in 1998, although research studies generally post-dated this period), and, to a lesser extent, on the role of the print and other media in relation to the economic crash of 2007–8. Fawcett's valuable study (2002), for example, concluded that:

> rhetorical and narrative structures *shape* and *constrain* the manner in which newspapers report conflict ... [and] those very structures can *position* newspapers as both politicians and storytellers in their coverage of conflict; this, in turn, means newspapers are likely to act strategically, finding the frame that best suits their purpose as politician and story-teller. In attempting to find ways of encouraging newspapers to promote 'win-win' as opposed to 'win-lose' frames of conflict, it will be necessary to address the power of these discursive structures, as well as the power of the political and professional cultures within which journalists operate.

Baker (2005: 386) highlighted a related problem, although the commercial realities of publishing, especially in print, suggest that it is not a problem with an easy solution. He concluded:

> Mainstream news channels certainly have greater power to distribute their versions to a greater number of people but they do so in competition with alternative sources of information and news. As is the case in Northern Ireland, alternative media might offer an absorbing insight into the political dynamics of conflict that is often lacking in mainstream coverage. The absence of alternative forms from the general analysis of news reduces our understanding of its role in the political process and makes it difficult to account for opposition to official and mainstream interpretations.

A later study on a narrower but still relevant theme, that of 'peace journalism', concluded that 'reporters are not innocents abroad, but complex decision-makers in an untidy world', and that 'even if one might agree with Peace journalists about any part of their diagnosis, their solutions are often the wrong ones' (Loyn, 2007: 57). Thomas Hanitzsch, in a more detailed comment, suggested that:

> The inherent logic of news production is another limitation of peace journalism. First, it is an unwarranted assumption that, given the

salience and importance of news values in public communication, that
peace journalism will prevail in a commercial media system that is driven
by market forces. Second, because of their specific functionality, some
media ... are more than others ... sympathetic to the ideas of peace
journalism. Third, it is difficult, if not impossible, to implement the
values of peace journalism in traditional news formats where space and
time constraints do not allow a detailed elaboration of backgrounds and
causes of violence as well as its consequences. Fourth, and last, the
demand for complexity reduction leads to the use of highly standardized
narrative schemes which are often not compatible to the demands of
peace journalism. (Hanitzsch, 2007: 172)

Not that all the news from the academic community was bad. A joint
Irish-British study (Dekavalla and Rafter, 2016: 45) looked back as well as
forwards when it concluded that the media coverage of Queen Elizabeth's
visit to the Republic of Ireland in 2011 could be characterized as:

emphasising the message embedded in the media event and minimising
the visibility of any voices of dissent. It may have been the persuasive
power of the media event that led to this homogeneous representation.
Or perhaps the possibility that after almost 20 years of peace establishment efforts – which the mainstream media have supported in their
different stages ... the journalistic coverage of Anglo-Irish relations is
gradually passing into a sphere of consensus ... where peace and concil-
iation are the only acceptable options for the future.

While the latter research studies focused exclusively on events in Northern
Ireland and the print media's response to them, the main burden of research
emphasis in the Republic during this period was in relation to the – much-
contested – role of the media in the period immediately preceding the
economic crash and the subsequent recession in the Republic from late 2007
onwards. The most relevant studies here are Cawley (2012), Fahy, O'Brien and
Poti (2010), Silke (2015) and Mercille (2015). Studies like these were the
background to the examination of the role of newspapers in the years before
the boom in property prices that was an intrinsic element of the crash; this
examination also took place as part of the Oireachtas banking inquiry in
March 2015. Although that inquiry's final report did not feature any findings
or recommendations into the role of the media, its report of the evidence

given contains much valuable material about the role of the media (Oireachtas, 2016).

There was an undercurrent of criticism of the print media after 2010 for having allegedly underpinned an unsustainable boom in house prices. Some of it was uttered by no less an authority than the former editor of the *Irish Times*, Conor Brady, who responded to some degree in the affirmative to his own questions: 'Was the forming of this crisis reportable earlier? Were emerging trends apparent? Did they [the news media] do as good a job as they might have in flagging the approaching storm?' (*IT*, 6 Mar. 2010).

The reality, however, was complex, and involved the sometimes confusing intermingling of different themes (particularly related to housing): whether print media coverage was too much beholden to powerful advertising interests and for this reason had accelerated the boom and therefore intensified the crisis; and whether press coverage post-crisis had been ideologically limited to supporting 'austerity' and failing to canvass other options because they lay outside a neo-liberal consensus.

It is difficult to come to any firm conclusion on the extent to which media coverage accelerated the boom, or indeed minimized the visibility of the approaching crisis. Most of the research focused on the reporting of the crash itself and of the government's attempts to manage it; to an extent, this research found or implied that the elements governing this coverage could also be discovered in the rear-view mirror for the pre-crash period. Although a number of them have identified some of the primary ideological characteristics of print financial journalism in particular during the period as a whole, none of them provided the evidential or empirical basis required to demonstrate an unambiguous causal link between these ideological characteristics and the outbreak, timing or magnitude of the crisis as a whole. Though Bourke (2008) and Fahy, O'Brien and Poti (2010) provide evidence of a certain subservience by a small number of financial journalists to powerful commercial interests pre-2007, the truth of this thesis has limited evidential value when it comes to answering Brady's questions above.

Although Brady's critique was strong, his editorship of the *Irish Times* ended in 2002, just as the economy began to recover from the brief dot-com crash of 2000–2. In the years between 2002 and the beginning of the crash, the available evidence suggests that the newspapers, rather than acting as cheerleaders for the runaway inflation of property values, were for the most part merely reporting what other, supposedly more authoritative, voices were predicting. In the two calendar years 2006 and 2007, there was approximately

one use of the term 'soft landing' per week in the *Irish Times*, but these references were in reports of statements by others (notably by the two taoisigh,
Bertie Ahern and Brian Cowen). The papers duly noted warnings about
what was happening from authorities such as the International Monetary
Fund and the OECD during this period, but no domestic expert was
reported as warning what might happen until Alan Ahearne, the economist
then working in Washington and later to head the state's Fiscal Advisory
Council, noted that 'the landing may not be as soft as the consensus view'
(*IT*, 28 Sept. 2007).

Much of the post-crisis academic work, in fact, was primarily devoted to
investigating the ideological and other biases of Irish economic and financial
journalism generally. Maher and O'Brien's approach was broad, combining
elements like religion, theatre and fiction as part of its scene-setting for the
crisis, although it eschewed any analysis of the fact that – with the exceptions
cited earlier – the principal prophets of doom were to be found in the
relatively unnoticed new media. Cawley's study, which was effectively based
on an analysis of the framing of reporting about the effects of the crisis in
the public and private sectors, concluded that:

> The economic crisis, as a multi-dimensional, inter-connected and highly
> technical process, placed considerable strain on the political-economic
> position of institutional actors in Ireland, and on the news-model trying
> to interpret and define the crisis for the public. Despite the dynamic
> nature of mass-media communication, where sponsored messages are
> continuously promoted, contested and in need of renewal, the news-
> model remained remarkably stable from July 2008 to June 2010 despite
> the unprecedented effects of the crisis on Ireland's economic, political
> and social systems. Division, opposition and competing economic and
> social interests endured as a key framing mechanism, weakening percep-
> tions of social cohesion and common interests. (2012: 613)

Mercille's two most powerful critiques (Mercille, 2014; 2015) dealt primarily
with coverage of the austerity, post-crisis years, and it is difficult to discern
any causal connection between this period and financial journalism in the
pre-crisis period. So did Titley (2013), although his research usefully drew
attention to the existence of a network of community activists, journalists
and academics prepared to explore ways of challenging the dominant framing
of the crisis.

By the time that the Oireachtas inquiry into the crisis was held in 2014 to 2016, however, the subtext apportioning blame to the media for inadequate vigilance had taken hold, and a number of senior editors and journalists gave evidence on the topic. Almost all the senior journalists concerned (with the exception of one who detailed his own experience of the influence of corporate interests on financial journalism generally) defended their records and asked, not unreasonably, how they might have been expected to predict, much less prevent, the economic crisis (and in particular its housing and construction components) when a myriad of authorities and regulators, up to and including the Central Bank – with infinitely greater expertise and resources – had not done any better. It may be that the principal implication for journalism is in the area of journalism education and training: French is eloquent on 'the confusion surrounding journalism education in general, not only in Ireland but in most European countries' and the fact that 'journalism has been slow to develop in higher education' (2007: 48).

Two years before the crash, in fact, the historian T. Ryle Dwyer, who was not an economist, wrote the first of a number of articles in the *Irish Examiner* warning about a possible dramatic downturn (*IE*, 5 May 2005). It was not until eighteen months later that the economist Professor Morgan Kelly, of University College Dublin, wrote:

> Pilots define a soft landing as one that you can walk away from. Looking at the price collapses in places like Finland and the Netherlands, and the building bust in Arizona, Ireland could be heading for something closer to what they call CDIT: controlled descent into terrain. You are happily descending through cloud, thinking yourself at a safe altitude, until suddenly you smack into a hillside. (Kelly, 2006a)

This passed relatively unnoticed, however, until later in the same year, when he advised readers of the *Irish Times*:

> If ... we look at what has happened to other small economies where sudden prosperity and easy credit drove house prices to absurd levels, we should be very worried indeed. If the experiences of economies similar to ours are anything to go by, we may be looking at large and prolonged falls in real house prices of the order of 40–50 per cent and a collapse of house-building activity ... In effect, the economy is based on building houses for all the people that have got jobs building houses.

Economists call this a multiplier-accelerator process and it is very
unstable. (Kelly, 2006b)

This article appeared three days after Christmas, which no doubt
contributed to its relative obscurity, but it was not entirely invisible: the
department of finance, which with many others was predicting a 'soft landing'
for the property market, was rumoured to have queried his analysis informally
with Professor Kelly's university employer. The *Irish Times* also came in for
some behind-the-scenes criticism. One way or another, however, his prognos-
tications – like the earlier ones in the *Irish Examiner,* failed to attract wider
attention, much less support, until considerably later.

The most detailed examination of the relationship between print media
coverage and the boom/bust phenomenon is that of Silke (2015). Although
he argued that it was very likely that the coverage of the housing issue
contributed to the housing bubble, he refused to adopt a simplistic analysis
that would have placed an inordinate amount of responsibility on the print
media, noting in particular that the de-staffing of the media industry (subsequently accelerated by the economic crisis) had 'a direct effect on media
practice and may even have impeded critique of the overall economic system
itself' (Silke, 2015: 320).

His critique of what he described as the failure by the newspapers to fulfil
'their normative watchdog role', however, is somewhat short on prescriptions
for improvement, other than the suggestion that 'the Irish communications
and journalist research community, though under-resourced, should make the
issue of journalism and power a priority' (Silke, 2015: 424–5). And, as he
points out, his analysis 'begs the question of what elements of autonomy
remain within the mainstream media sphere?' (Silke, 2015: 325). While the
research community of which Silke is such an excellent example can powerfully raise such questions, finding answers – in an industry that traditionally
for the most part reflects the dominant, profit-driven characteristics of the
society and economy within which it operates – is quite another matter.
There was some evidence of a fight-back by unions: threatened staff reduc-
tions at Alpha Newspapers in Northern Ireland (owned by the unionist peer
and former cabinet minister Lord Kilclooney) resulted in an increase from 25
per cent to 85 per cent in the membership of the National Union of
Journalists at these titles, and employees won an average 30 per cent pay rise
as well as an incremental salary scale and union recognition. Media ownership
itself remained a vibrant topic in the Republic. A report commissioned by

the Sinn Féin MEP Lynn Boylan and prepared by lawyers for the European parliament grouping GUE/NGL (the European United Left/Nordic Green Left) of which she is a member claimed that the dominant position of Denis O'Brien in the Irish media landscape meant that Ireland had 'one of the most concentrated media markets of any democracy', and that this ownership pattern, together with Irelands libel laws, created a 'perfect storm which threatens media plurality and undermines the media's ability to perform its watchdog function' (European Parliament, 2016). This prompted a vigorous riposte from O'Brien, in which he expressed his belief that some media companies would not survive past 2020 without radical restructuring, including substantial funding, and added: 'I do not believe the Irish media is objective in relation to matters relating to itself. The prime reason is survival. Every media and journalist knows that the future of traditional media is bleak' (*SBP,* 26 Oct. 2016).

The year 2016 saw an outbreak of inter-media hostility. There was a degree of public rivalry between media companies, and overt or implied criticism of each other embedded in detailed reporting of each other's difficulties, of a kind rarely seen before, even during the heyday of Tony O'Reilly's hegemony at Independent News and Media. Many nerves remain on edge, and the future may be bleak or uncertain, but there are increasingly focused and experimental searches for new solutions.

11 From legacy to hybrid media, 2008–16

As already noted, the period from 2008 was to be the most seismic so far in the long history of Irish print and broadcast media. As detailed below much of this stemmed from the impact of the post-2008 economic crash. Although in other parts of the world, print circulation in particular had begun a precipitous decline much earlier, the Celtic Tiger boom had delayed a similar deterioration in Ireland. However, the recession also coincided with a maturation of the online media sector. Though routinely identified as potentially undermining older media forms from the mid-1990s, it was not until the 2000s that a combination of widespread broadband connectivity, the spread of genuinely mobile internet access via smartphones and the upsurge in social media use rendered online outlets ubiquitous.

As with the arrival of radio and television in the twentieth century, older forms often responded to telecommunications-infrastructure-based upstart forms in a defensive manner, although to a greater or lesser extent established (often referred to as 'legacy') Irish media brands would increasingly recognize that the internet created an opportunity to extend those brands into new markets. By 2013, Andrew Chadwick would argue that it made less and less sense to divide media along 'old' and 'new' lines, instead emphasizing the inter-relationships of and convergence between such media to explain both how 'newer media practices in the interpenetrated fields of media and politics adapt and integrate the logics of older media practices', and conversely how older media practices adapted/integrated newer logics.

From 2008, the Irish media were increasingly replete with such interpenetrations and increasingly came to resemble what Chadwick (2013, 44) characterized as 'hybrid media' forms. Yet the creation of the underlying socio-technical infrastructure on which these hybrid media formations were built was not without its growing pains.

For most of the twentieth century, Irish telecommunications had been in catch-up mode. By the start of the 1980s, fewer than 50 per cent of homes had a basic landline connection and 100,000 more languished on a waiting list. Telecom Éireann – a newly created statutory body – had formally taken over the telecommunications service in January 1984. A five-year £650 million capital-development programme – more money than had been spent in the previous history of the state – would result in a remarkable transformation: by the end of the decade Telecom Éireann had not merely ended the waiting lists but had, after adding mobile telephony and data services to its offerings, begun to produce substantial surpluses. This coincided with intense pressure from the European Commission to liberalize the telecoms market. In 1990, Denis O'Brien, then best-known as the owner of a Dublin commercial radio licence, had established Esat Telecommunications, which, in 1993 was licensed to resell capacity on lines leased from Telecom Éireann to business customers. Two years later, Esat, in partnership with the Norwegian company Telenor, acquired the first private licence to offer mobile-telephone services, and, with the 1999 acquisitions of internet service providers (ISPs) EUnet and Ireland On-Line, became for a period the biggest ISP in Ireland.

EUnet had been established (as Ieunet) at Trinity College Dublin in 1991, but it was Ireland-On-Line's (IOL) launch the following year that inaugurated the first direct-to-customer ISP. In 1997 Telecom Éireann would both establish its first ISP, TINET, and acquire an existing ISP, Indigo, which it continued to run as a separate operation. Thus, as in other telecoms sectors, at the turn of the twenty-first century, Telecom Éireann and Esat dominated the Irish internet-service-provision market.

The initial offerings were not conducive to prolonged periods online. Connection speeds were slow and costs were high. There was little incentive to engage in extensive browsing, which tended to be the preserve of those with internet access at work. Figures from the office for the director of telecommunications regulation (ODTR) in 2002 suggested that the average user spent just 218 minutes per month online. By contrast, television viewers spent a roughly similarly duration in front of their sets each day (ODTR, 2002a).

Household internet penetration would rise from 5 per cent in 1998 to 20.5 per cent in 2000 and to 33.5 per cent in 2003, but, as late as 2003, 85 per cent of subscribers remained on dial-up connections and fewer than one in five domestic users went online on a daily basis. One 2003 report suggested that many consumers (especially those over 45 years of age) 'simply have no compelling reason to go online', because it was not required for their work, and, in a pre-social-media age, because 'their friends are not online' (Amárach, 2003). Nonetheless, household internet penetration continued to grow. The number of narrowband internet connections doubled from 100,000 to 200,000 in the nine months after autumn 2003, and household penetration grew from 38.2 per cent in 2004 to 45.1 per cent in 2005 (ODTR, 2006: 19).

In 2002 broadband connections were generally limited to business customers, who accounted for most of the 3,000–4,000 subscribers using high-speed connections on leased lines or lines connected to one of the just thirty-five DSL-equipped exchanges (all of which served Dublin). This figure was low – equivalent to just 0.3 per cent of all potential residential and business customers. These early broadband markets were dominated by fixed-line telecoms operators Eircom and, to a much smaller extent, Esat (by now partnered with British Telecom).

The ODTR was not optimistic about the pace of broadband rollout – having been privatized via a stock-market launch in 1999, most of the shares in Telecom Éireann (or 'Eircom', as it had been rebranded in advance of privatization) had been acquired by the private Valentia consortium by 2002. The new owners made it clear that they were looking to cut costs rather than embark on the kind of capital investment a rapid national expansion of DSL (digital subscriber line) technology would demand. Cable infrastructures presented the greatest potential competition to Eircom, but the ODTR noted that – post-dot-com boom – neither of the existing market leaders, NTL and Chorus, had access to the funding needed to finance the infrastructure upgrade. Visions of offering cable-based 'triple-play' (television, phone and broadband) packages were receding from view, replaced by an effort to catch up with Sky's remarkable (245,000 Irish subscribers by summer 2002) dominance of the digital television market (ODTR, 2002b).

Thus, in practice, it was Eircom that established first-mover advantage in broadband as it gradually expanded the number of DSL-capable exchanges. The number of DSL subscribers grew from 7,350 in mid-2003 to 125,000 by the start of 2005, equivalent to 85 per cent of all broadband subscribers. The year 2005 proved a tipping point in terms of narrowband versus broadband subscriptions: the latter grew from 29 per cent to 42 per cent of all subscriptions over the year and broadband subscriptions would outstrip narrowband for the first time in the last quarter of 2006. Indeed, Ireland's broadband-connection growth rate between 2006 and 2007 was the highest in the OECD and, by the end of 2011, narrowband subscriptions had all but disappeared, accounting for less than 2 per cent of all connections. This was reflected at household level: just 3 per cent of Irish households had broadband in 2004. Thereafter, domestic penetration rates doubled each year, to 7 per cent in 2005, 13 per cent in 2006, 31 per cent in 2007 and 63 per cent in 2008. Even so, Ireland remained something of a laggard by EU standards: as late as June 2008, the 8.8 per cent per capita broadband-penetration rate was well behind the 14 per cent EU average.

A key factor in driving both the increase in overall internet subscriptions and also the move to broadband had been the entry of mobile-phone operators (Vodafone in November 2006, O2 in summer 2007) into the mobile-broadband market, offering users access to wireless broadband via USB dongles plugged into their PCs. This was a particular boon to those – often rural – customers who though living beyond the reach of Eircom's DSL exchanges and the main cable-company infrastructures were served by the connectivity umbrella offered by third-generation mobile-phone networks (see below). By autumn 2007, the 88,000 mobile-broadband subscribers already accounted for more than 10 per cent of all broadband connections and this figure grew to 300,000 by the end of 2008, equivalent to a quarter of the 1.2 million total broadband customers. This in turn undermined Eircom's dominance of the – by now increasingly synonymous – internet and broadband markets. Having accounted for the bulk of broadband subscriptions since the beginning of the decade, its share of the broadband market fell below 50 per cent for the first time at the end of 2009, as the mobile-broadband providers saw their share rise to 32.4 per cent (Artero, forthcoming).

Total internet subscriptions reached just under 1.7 million by the end of 2011, representing the mature phase of commercially driven expansion of fixed-line and mobile broadband connections. For the next five years, although there were significant variations in the market shares of various connectivity modes, the overall number of connections remained fairly static. Mobile broadband's market share (dominated by mobile-phone operators Vodafone, O2, Three and Eircom) peaked in early 2011 at nearly 600,000, but subsequently fell back to less than 400,000. A key issue here was the inherent technical limitations of mobile broadband's reliance on third- and fourth-generation mobile-phone infrastructures, most notably in terms of download speeds. Although broadband implied greater data-connection speeds, these were defined in relation to other data-connection technologies: as late as 2007, for the majority (59 per cent) of residential subscribers, broadband meant sub-2mbps connections.

In that context, the download speeds offered by 3G-based mobile broadband were at least comparable with speeds available via wired connections. However, the marked increase in the number of cable-broadband subscribers after 2009, as UPC/Virgin ramped up its investment, was reflected in a similarly marked increase in average connection speeds – the percentage of customers on sub-2mbps speeds fell from 47.3 per cent at the beginning of 2008 to 24.2 per cent by the end of 2009. By contrast, those with 2mbps-plus connection speeds rose from 52.7 per cent to 75.8 per cent over the same period. Thereafter, a 'connection-speed race' ensued between cable, mobile broadband and fibre-based providers with, by early 2017, claims of 340mbps (Virgin), 155mbs (Vodafone Mobile) and 1gbps (Eir) services highlighted in promotional literature. These appear to have reshaped the disposition of market shares: by 2016, Eir (as Eircom had been rebranded in September 2015), accounted for about 35 per cent of all subscriptions, while Virgin, which in its earlier guises as NTL/Chorus/UPC had lagged well behind the rest of the market, surged forward to account for 28 per cent of all broadband subscribers.

As of 2016, Ireland's 1.7 million connections represented just over 52 per cent of business and domestic premises in Ireland. Although some research suggested that 67 per cent of homes had a broadband connection by 2014 (Amárach, 2014) there remained a significant gap in broadband access for those living in rural areas, where laying wired infrastructure was considered uneconomic. The launch of 4G-based mobile broadband services was initially identified as potentially addressing these gaps, but as early licence require-ments limited rollout targets to 70 per cent of the population, 4G network upgrades were initially concentrated in high-population-density areas. Notwithstanding repeated commitments by successive governments to roll out a national broadband strategy, practical action was slow. After the 2016 general election a new national strategy was unveiled that was designed to bring broadband to an additional 300,000 rural homes by 2018, which would bring the country to 80 per cent household penetration. It did not foresee the achievement of 100 per cent until 2020 at the earliest.

This emphasis on fixed broadband or mobile broadband to some extent obscures the transformation in connectivity following the post-2008 explosion in smartphone ownership. Mobile phone services had first been launched in Ireland by Eircell, a Telecom Éireann subsidary, in 1986. The high cost of sets and calls together with the limited reach of the network meant that initial growth was slow: there were just 13,800 Eircell subscribers by 1990 (Telecom Éireann, 1990). However, with the introduction of the first digital GSM (or 2G) service in 1993, subscription rates began to increase, and the 100,000-subscriber mark was breached in 1995. In that same year, the first private mobile-phone operator, Esat Digifone, was licensed by the state. By the time a third operator – Meteor – began operations in 2001, Eircom had sold off its mobile subsidiary to international giant Vodafone. (Esat would later be sold to British Telecom, which in turn hived it off as the O2 network to Spanish telco Telefonica in 2005. The same year saw Eircom acquire Meteor as it sought to re-enter the mobile market.) Competition drove an enormous expansion in subscription rates. Having reached nearly 45 per cent of the total population by the beginning of 2000, mobile penetration leapt to over 70 per cent by January 2002. By June the same year there were just under 3 million subscribers in the country (ODTR, 2002a).

The limited data-carrying capacity of GSM networks (typically 9.6kbps) meant that voice calls constituted the bulk of traffic (and thus revenue) on these networks. Data use was limited to text messaging, a particularly popular mode of communications with younger user cohorts. The introduction of WAP (wireless application protocol) services in 1999 promised to unlock cyberspace for mobile users, but in practice the WAP's severely cut-down versions of HTML sites did not incentivize widespread use. However, the arrival of a new standard, 3G or UMTS (universal mobile telecommunications services), first made available in the Irish market by Vodafone in November 2004, transformed data rates.

In July 2005, a fourth, 3G-only network – 'Three' – was launched in Ireland by Hong Kong-based Hutchison Whampoa. The presence of four networks drove further price competition, and by the end of 2005, the number of subscriptions (4.2 million) outstripped the size of the population (4.1 million) for the first time. An Amárach Consultancy report in 2005 found that of the 85 per cent of residential users with a mobile phone had two or more subscriptions, mainly to distinguish between personal and business use. Yet this still placed Ireland towards the bottom rank of EU-15 countries, and behind the EU average whereby mobile-phone subscriptions were equivalent to 107 per cent of the population (ODTR, 2006).

If the arrival of 3G – the connection speeds of which were subsequently (2013) surpassed by the rollout of 4G services offering up to 25mbps – put the underlying infrastructure for mobile connectivity in place, the subsequent arrival of smartphones transformed mobile subscribers from occasional visitors into near-permanent denizens of the online world. Although Canadian firm Research in Motion (RIM) had introduced the Blackberry smartphone in 2003, this was primarily aimed at a business market and emphasized email access over richer web experiences. By contrast Apple's launch of the iPhone in summer 2007, with its touch-screen interface, offered a more seductive gateway to online connectivity. O2 Ireland began offering iPhone handsets in March 2008, and other networks gradually followed suit. Other smartphone operating systems quickly emerged to eat into the market share of Apple's iOS, most notably Google's Android, an open-source operating system, which, following its 2008 release, was quickly exploited by a range of hardware manufacturers, including Samsung and, for a period, Google itself via its Nexus range.

The smartphone proved an instant success with Irish users. Having introduced the iPhone in June 2010, Vodafone reported that 50 per cent of all devices sold by the year's end were smartphones (ComReg, 2010). Though 3G SIM cards were not exclusively used in conjunction with smartphones, in autumn 2011, ComReg – successor to the ODTR – concluded that of 4 million 3G subscriptions (from the total 5.5 million mobile-phone subscriptions in Ireland) about 1.9 million were 'active'. In other words 34.5 per cent of all mobile-phone subscribers were already extensively making use of smartphones to go online (ComReg, 2011). Surveys conducted by Eir suggested that 50 per cent of adults with a phone had a smartphone by April 2013, increasing to 64 per cent by August the following year (Eir 2013a, 2014b). By the close of 2016, ComReg reported that smartphones accounted for 88 per cent of all handsets (ComReg, 2016).

There was no doubting the speed with which smartphones (and to a lesser extent tablets) became integrated in daily lives. By 2013, 38 per cent of Irish smartphone and tablet users asserted that they could no longer 'live without' those technologies for accessing online information; 20 per cent considered them essential for accessing print media online. The extent of reliance varied according to age cohort: 55 per cent of 16–24 year olds argued that they 'could not live without' their smartphone (at all), as compared with 0 per cent of over 65s. Such attitudes mirrored ownership – by September 2013, 81 per cent of Irish 16–24 year-olds owned a smartphone as opposed to 13 per cent of over 65s. Indeed, 75 per cent of the latter cohort conceded that they simply didn't understand computers and new technology, as compared with 7 per cent of under-24s (Eir, 2013b).

Some users were almost literally joined at the hip to their devices. Research commissioned by Eir in 2013 suggested that in addition to 'normal' locations for online access, 71 per cent of Irish users consulted their smartphones in the car, 51 per cent on public transport and 27 per cent admitted to using them while on the toilet (Eir, 2013a: 24). This extended to time spent outside Ireland: 43 per cent of smartphone and tablet users accessed – mainly Irish – news apps while travelling overseas by the latter part of 2013 (Eir, 2013b: 34). By the end of 2014, more people identified smartphones as the one media technology they could not survive without than any other (Eir, 2014b). (Of those who owned one, 52 per cent professed to be unable to exist without it for a week.)

Taken in conjunction with the figures for fixed and mobile broadband connections, the extensive diffusion of the smartphone pointed to a sea change in social organization, including media use. The economy, relationships and the nature of the public sphere were increasingly in flux due to developments online. By 2016, online spending in Ireland was worth about €7.5 billion, and the digital economy as a whole accounted for 6 per cent of gross domestic product (GDP) (Virgin, 2016: 3). Relationships broadened to include social-media friendships – one 2015 survey suggested that the average Irish 16–35-year-old maintained 360 friendships on Facebook (Thinkhouse, 2015). And, as discussed below, online media – whether net natives, web versions of legacy outlets or accessed via social media – gradually surpassed radio and print in terms of popularity, to a point where, by 2016, only television could vie with it for the status of most influential news medium.

For all this, the internet in some respects proved less of a threat to legacy media institutions than might have been anticipated. The nature of the smartphone in particular – small and portable – meant that media outlets, whether native to print, broadcast or online, were – network permitting – available at any time in any place. Thus, in terms of time spent consuming media at least, online media use proved more additive than cannibalistic, filling in the gaps in what had hitherto been regarded as 'empty time' (Siapera, 2012: 164) rather than taking time away from traditional media consumption.

Irish rates of social-media use were high by international standards. A 2011 Eurobarometer study found that 68 per cent of Irish internet users described themselves as using social media, well above the EU average of 52 per cent. Notably, the same survey found that Irish internet users were particularly active in content creation: 15 per cent described themselves as maintaining a regular blog, three times the EU average and higher than any other EU-27 country bar Cyprus (Eurobarometer, 2011).

Facebook stood out as the most popular site: the platform's own figures pointed to an increase from 200,000 to 2.5 million unique Irish monthly visitors between October 2008 and October 2015 (*IT*, 10 Sept. 2009). Having an account and using it were not the same thing, however. Although a Red C poll in 2011 suggested that 76 per cent of Irish adults had a Facebook account by 2011, when Ipsos/MRBI began routine tracking of social-media use the same year, they found that just 50 per cent of Irish adults were maintaining their accounts. This would subsequently grow to 59 per cent by the start of 2014 and to a 67 per cent peak by October 2016 (it subsequently fell back to 64 per cent by January 2017). Other social media also acquired significant user bases in Ireland over the same period. Twitter climbed from 7 per cent of all adults in 2011 to 31 per cent in January 2016; LinkedIn from 9 per cent to 28 per cent over the same period. Few matched Facebook's capacity to draw users back on a routine basis, however: 74 per cent of Facebook users attended to the site on a daily basis by January 2017, as compared with 56 per cent of the users of Instagram (by then the second most commonly used social medium in Ireland).

Growing use of social media contributed to the growth in time spent online. An Amárach survey in 2007 estimated that Irish adults spent an average of fifty minutes a day on the internet. Though not using the same methodology, a Red C poll in 2011 suggested that this increased to 120 minutes in 2009 and to 164 minutes in 2011, a rise that coincided with the first wave of smartphone diffusion (Red C, 2011: 22). Yet at the same time, television viewing levels – in terms of minutes watched per day – also increased as the twenty-first century progressed, reaching 189 daily minutes per capita by August 2014 (Eir, 2014b).

It was less clear whether television continued to garner the same level of attention, however, as the phenomenon of 'multiscreening' or 'media stacking' took hold. Eir found that 76 per cent of 16–24-year-olds often Facebooked or Tweeted while watching television (2013b: 30), and a 2014 Amárach survey suggested that 79 per cent of all adults were at least occasionally simultaneously watching TV and using a laptop, smartphone or tablet.

Yet, even when online, users were not necessarily ignoring legacy media institutions. Although Facebook emerged as the most popular Irish smartphone app by 2013 in one survey (75 per cent of respondents acknowledged they had downloaded it), the RTÉ News Now app emerged as third most popular, appearing on 56 per cent of smartphone-users' screens (thus making it as common as YouTube) (Eir, 2013b: 36). Although Facebook trounced all opposition in the 'most addictive app' category – with 45 per cent of respondents identifying it as such – RTÉ at least made the top five among all user cohorts.

If elements of legacy media were able to maintain the absolute size of their audience, they would prove less successful when it came to the battle for users' pockets, or rather for the pockets of those advertisers who fought for consumer attention. The capacity of social media and search engines to scrape the personal data and preferences of individuals online yielded vast sets of 'big data', which allowed those same online players to construct hitherto impossibly specific audiences attuned to the particular cohorts sought by advertisers. That these audiences were disproportionately accounted for by younger cohorts (by 2011, 18–24-year-olds spent 214 minutes a day online, as compared with the 122 minutes expended by 45–54-year-olds), who were notoriously (and increasingly) difficult to access through traditional forms of advertising, made them all the more attractive, and thus lucrative. In sum then, until 2008, the combination of the overall health of the economy together with the still nascent status of online media made it possible for legacy media to treat online counterparts as relatively distant threats. From 2008 on, however, the online context was an omnipresent factor in virtually every decision the more established media made, as the following pages will demonstrate.

Although gloomy economic forecasts were already increasingly prevalent, the annual reports for media firms published in spring 2008 (but relating to 2007) gave a false impression of a broadcasting industry in rude health. Based on a €9.49 million turnover, the €2.94 million in pre-tax profits earned by UTV's FM104 up to March 2007 made it the most profitable local-radio station in the country. In that still-optimistic period, the station projected that profits for the coming year would increase to €4.6 million. Communicorp's Today FM posted profits of €7.4 million on €19.4 million turnover, while TV3 saw ad revenues increase 10 per cent to €62 million. And, in March 2008, buoyed by a 2007 surplus of €26.4 million, RTÉ announced plans for Diaspora TV, a satellite service offering RTÉ content, targeting Irish emigrants in the UK.

Yet, the good times were coming to an abrupt conclusion. As the property bubble supporting the latter years of the Celtic Tiger burst, August 2008 saw RTÉ Director General Cathal Goan warn of the 'most financially challenging' circumstances RTÉ had experienced for some years. As the economy entered freefall, television and radio advertising spend would drop by 31 per cent (from €414 million to €286 million) between 2007 and 2013 (Oliver and Ohlbaum, 2013: 23). Though this decline would impact all stations – commercial and public – the long recession inflicted a double hit on RTÉ as state-led austerity measures cut its licence-fee revenues by more than 10 per cent after 2008.

Initial RTÉ estimates for 2008 anticipated an €11 million surplus. By November that year, the most optimistic forecast was for break-even, if the station could find €27 million in cuts. This pattern recurred over the next five years: the station would note a revenue shortfall and identify cost savings, only to encounter a further shortfall. The savings came through a number of means: first pay freezes, then pay cuts for all staff (ranging from 5 per cent to 17.5 per cent per cent, depending on salary), including high-profile on-screen talent. These were insufficient to stave off huge losses. The cumulative deficit of €49 million recorded between 2009 and 2011 ballooned to €65 million for 2012 alone. Much of this resulted from the short-term cost of executing a longer-term cost-saving strategy: reducing staff numbers through voluntary redundancies. As 500 staff departed between 2008 and 2012 (more than 20 per cent of the total), 'restructuring expenses' rose to €46 million in 2012.

Inevitably, the effects of the cuts trickled down to the schedules. Expenditure on in-house productions fell from €245 million in 2008 to €209 million in 2009, affecting all areas – drama, factual and news and current affairs. This had a knock-on impact on the independent production sector. Having spent €79 million on such commissions in 2007, RTÉ cut this to €54 million in 2010.

Cuts could only go so far, though, and RTÉ explored other ways of raising revenues. On his appointment as head of the RTÉ Authority in February 2009, PR consultant Tom Savage immediately argued for an increase in the licence fee. However, having fought to tie the fee to inflation, RTÉ was now in a difficult situation: as the economy stagnated, the Consumer Price Index actually dropped, undermining any request for a fee increase. Indeed, if anything, the state's actions worsened the situation. Having replaced Cathal Goan as director general in November 2010, former producer Noel Curran was confronted with the financial fallout of the December 2010 national budget: €10 million was diverted from the licence fee directly to TG4, while an additional €4.3 million was lost to an enhanced Sound and Vision fund. In addition to these revenue losses, RTÉ (along with commercial broad-casters) faced a new €2.5 million regulatory levy, while the department of social welfare announced a reduction in its subvention of licence fees for social-welfare recipients. The net cost of these measures to RTÉ was €20 million. And, to add insult to injury, Curran faced additional unexpected costs incurred by the need to cover the visits of UK monarch Queen Elizabeth II and US President Barack Obama in May 2011.

Other money-raising measures were pursued. Abandoning its internal sponsorship guidelines separating news and current-affairs programming from sponsorship, in 2009 RTÉ announced a nine-month deal with HSBC Corporate Banking to sponsor Radio 1's *Drivetime*. Extensive efforts were also made to mitigate losses stemming from high-profile presenter shifts. Following Pat Kenny's March 2009 announcement that he was stepping down from presenting the *Late Late Show*, the decision to appoint the relatively youthful Ryan Tubridy as the new host was clearly aimed at lowering the audience's age profile to one more attractive to advertisers. Nonetheless, a gradual ratings decline ensued, the show's audience falling to less than half a million viewers by early 2017. Though individual editions bucked the trend – in 2014, the annual 'Toy Show' secured nearly 1.6 million viewers – the format seemed less and less relevant to a changing society.

Tubridy was also deployed to raise the age profile of the ailing 2FM. Despite an ebbing market share (from 15 per cent in 2004 to 12.8 per cent by 2008), 2FM had returned a net profit on its operations during the boom. However, with the recession, mounting losses ensued as 2FM's share of the national market halved to less than 6 per cent by the close of 2016. The station found it particularly hard to recover from the tragic death of Gerry Ryan, by far its most popular presenter, from cardiac failure in May 2010. It fell to Ryan Tubridy to try to retain his charismatic predecessor's ratings, and the associated €4–5 million in annual revenues.

The recession did not exclusively affect RTÉ, although the scale of its impact elsewhere varied enormously. Some hint of declining market confidence emerged after five independently owned local stations – Ocean FM, Mid-West Radio, KCLR, KFM and Tipp FM – were offered for sale as a single block in April 2008. Scottish investor Richard Findlay initially offered €50 million for all five, but as the recession radically altered radio-market prospects, he reduced this to €36 million by November 2008 and the deal foundered. Matters hardly improved in the half decade that followed. When, in September 2012, the BAI called for applications for twelve local-radio licences due to expire the following year, no new commercial entrants emerged to challenge the incumbents. Those new players like north-west regional station i102–104 and the 'multi-city' 4FM that came on air after 2008 also struggled. In March 2010, i102–104's parent announced the closure of its Galway studios, making twelve staff redundant in the process. A year later, the operations of i102–104 and i105–107 were merged to create a single northern region franchise under the name iRadio.

iRadio at least remained on air. The same was not true of the City Channel cable-television stations which, despite the deep pockets of their owner,

Liberty Global, went to the wall in September 2011. They had not made a profit since their 2005 launch, but revenues collapsed by 50 per cent between 2009 and 2010. A similar fate had threatened Channel 6 by early 2008. By the close of 2007, even before the recession hit, Channel 6 had accumulated losses of €11.4 million, and the directors' statement relating to the 2007 accounts expressed uncertainty as to the channel's short-term viability. A white knight was sought, and when Liberty Global baulked, the station turned instead to TV3, to which it had outsourced its advertising sales in January 2008. By July that year a deal, reportedly worth €2–3 million – saw TV3 absorb Channel 6, relocate it to TV3's Ballymount studios and, in January 2009, rebrand it as '3e'.

UTV faced falls in revenue in both the Republic of Ireland and, later, the UK. In November 2008, in anticipation of declining profits, it wrote off €26.5 million from the value of its Republic of Ireland radio stations. Revenues at the stations dropped nearly 20 per cent between 2008 and 2009, prompting a more-or-less ongoing series of cost-cutting exercises across the group. By 2012, as recessionary contagion spread to the UK, UTV media as a whole reported a drop in revenues and profits across Northern and Southern Irish operations. This remained particularly acute in the Republic. Though the group's overall 2012 broadcast-advertising revenues dropped 7 per cent, the fall was 12 per cent south of the border.

The recession also hit broadcast-news providers. In October 2009, Independent Network News, which had provided news content to the majority of local commercial radio stations since the 1980s, abruptly announced its closure on grounds of non-viability. Unable to fulfil their licence obligations without such content, the eighteen INN shareholder stations immediately advertised for a replacement. Bidders included UTV, but it was Communicorp's Newstalk that emerged victorious. If this raised concerns regarding concentration of editorial influence within Communicorp (although the stations serviced continued to provide their own local news content), subsequent events hardly improved matters. By October 2014, Newstalk was effectively serving the entire independent radio network with national and international news.

Communicorp had already taken measures to address the revenue decline at its Irish stations. Having raised €10 million in May 2008 by selling Donegal station Highland Radio, the following month saw Today FM and Newstalk move to a shared building – Marconi House – on Digges Lane. With Newstalk carrying accumulated losses of €19.1 million by 2007, further savings were necessary. In November 2008, the two stations pooled reporters to achieve cost efficiencies, and a month later Today FM announced job losses across all departments and the dropping of three shows – the Irish-language *Splanc*, *Culture Shock* and *Late Night Live*. This raised the ire of the BCI, which insisted that the two culture-oriented shows be reinstated.

In December 2008, Communicorp axed 140 jobs, equivalent to 10 per cent of its total workforce, and although in February 2009 Today FM Chief Executive Willie O'Reilly reassured his top stars that the station would not emulate RTÉ by seeking pay cuts, a pay freeze was nonetheless implemented. Within the Communicorp parent, July 2009 saw pay cuts of 5 per cent to 10 per cent across the board.

At the same time, Communicorp would extend a lifeline to Dublin-based Phantom Radio. Having gone on air in 2006, by October 2009 it had accumulated losses of €2.3 million, and in July 2010 Principle Management, a key investor, brought in Communicorp to address cash-flow problems. Communicorp took a 34 per cent stake and moved the station to Marconi House – although Communicorp insisted that this did not imply that they had operational control of the station. This merely delayed its demise however. A low market share (0.7 per cent by the end of 2012) would see a February 2014 rebrand as 'TXFM' (echoing Today FM), but also redundancies in a bid to make the station viable. Despite offering a genuine indy alternative to the pop-centric offerings of competitors, audiences remained low, and in April 2016 Communicorp declined to renew the licence for the station, which went off air in October 2016.

TV3 arguably weathered the recessionary storm better than most, though it was far from unscathed. Two rounds of – initially voluntary, then compulsory – redundancies between October 2008 and March 2009 saw the workforce reduced by fifty. The remaining 200 employees faced a 7 per cent pay cut (15 per cent at executive level). By the end of 2008, as profits halved to €10 million, TV3 owner Doughty Hanson wrote off €54 million of its investment in the state. The following year, a further €119 million in goodwill value – the amount by which a company's book value was enhanced by confidence in its future prospects – was written off.

In September 2008, the channel signed a new ten-year contract with the BCI. The new licence stipulated that the channel source 30 per cent of its output (measured in hours broadcast) from indigenous sources, up from the 25 per cent figure that had prevailed since 1998. For a channel previously reliant on acquired programming, this may have seemed like an unacceptable burden. Yet, under Doughty Hanson, it had embraced a new strategy of building audiences both by acquiring local sports rights but also investing in domestic-content production. In 2008, RTÉ's monopoly on live rights to GAA games came to an end as TV3 won the rights to screen 20 per cent of televised championship games. Under Director of Programmes Ben Frow, TV3 went from producing just four Irish shows in 2006 to 160 in 2008. Launching the channel's autumn schedule in August 2009, Chief Executive David McRedmond noted that the coming year would see 1,900 hours of home-produced content, 700 more than in 2006, including something never before seen on TV3: a multipart in-house drama, *The Guards.* It was followed in 2011 by the soap-style *Deception,* a drama for a post-Celtic Tiger Ireland, with a €1.2 million budget, half of which came from the Sound and Vision fund.

For the most part, Frow's schedule relied on a combination of UK soaps and talent shows, and Irish versions of international reality formats. It worked. By autumn 2009, TV3 had overtaken RTÉ2 to become the second most-watched channel in Ireland, and even if revenues were down, the channel remained profitable, returning €2 million in 2009. The €600,000 spent in 2010 on *Take Me Out Ireland* (a culinary-themed format originated by Fremantle Media in Australia) secured more than 300,000 viewers, despite being scheduled against *The Late Late Show* on a Friday night. Indeed, the TV3 show attracted a greater number of advertiser-friendly under-35s than RTÉ's light-entertainment behemoth. This, together with *The Apprentice Ireland* and a simulcast of ITV's *The X-Factor,* saw TV3 win 40–50 per cent of the ratings on Monday and Saturday, making it the most-watched Irish channel among younger audiences for much of autumn 2010. TV3's market share for 2010 rose to 13.3 per cent (from 12.9 per cent in 2009), to which 3e added a further 1.2 per cent gloss. Meanwhile, RTÉ2's share for the year receded further to 8.5 per cent, although RTÉ1 retained an unassailable lead with a 25 per cent share.

By the end of 2010, TV3's increasing emphasis on content production saw it planning to double the scale of its West Dublin facility, building Ireland's first high-definition television studio. This placed it ahead of RTÉ, which had already acknowledged that the existing Montrose campus could not accommodate emerging technologies such as HD. In autumn 2009, RTÉ unveiled Project 2025, a plan to gradually migrate operations to a new 500,000-square-foot complex at the northern end of the campus. Though planning permission was granted in 2010, RTÉ conceded that a lack of resources meant it was six to eight years away from launching full HD channels (although, in practice, RTÉ1 went high definition by Christmas 2013).

Yet even TV3 recorded substantial losses in some years. Tullamore Beta, the company operating TV3 and 3e, lost €6.7 million in 2011 and €7 million in 2012. All broadcasters watched nervously as the overall value of broadcast-advertising spend continued to decline. This sensitivity was made all the more acute by the recognition that new media were bucking that trend of decline. In 2006, some estimates suggested that online accounted for perhaps €42 million or just 3 per cent of total Irish advertising media spend. A 2008 report from consultancy Aecom suggested that online ad-spend had increased to €105 million or 8 per cent, and by 2009, when more definitive figures emerged from the Internet Auditing Bureau, it was obvious that online ad-spend was increasing at a remarkable rate. UK consultants Oliver and Ohlbaum estimated 18 per cent annual compound growth between 2007 and 2013 (2014: 23). Worryingly for legacy media, in July 2010, PWC noted that the online-advertising spending was shifting away from display adverts (i.e. those attracted by the websites of print and broadcast media) and towards 'paid search' advertising (i.e. adverts placed within the search engine results). Within the online-advertising market, print and broadcast media seemed to be losing out to Google et al.

These concerns were amplified by a technology designed to facilitate television viewing. By 2009 'virtual' personal video recorders (PVRs) permitting the recording of whole series at the click of a remote control button had become standard elements of the set-top boxes offered by television platform operators. By September 2013, two-thirds of those watching such 'time-shifted' television were skipping past the advertising that indirectly financed it, a figure that leapt to 86 per cent among 16–24-year-olds (Eir, 2013b: 28). Time-shifting remained exceptional, but it was inexorably increasing, rising from 8 per cent of all TV viewing in September 2012 to 11 per cent by 2016. Disconcertingly, time-shifting was most common during the peak hours, when television advertising was at its most expensive.

Given this, broadcasters increasingly contemplated other means of raising revenues. Some of these would require the approval of the broadcasting regulator. In March 2008, at the inaugural Independent Broadcasters of Ireland conference, then Communicorp Chair Lucy Gaffney had asked if the mooted replacement for the BCI would 'let us make money'. The Broadcasting Authority of Ireland, which emerged in September 2009 after the passage of the 2009 Broadcasting Act, appeared to answer Gaffney's question in the affirmative. From July 2010, the BAI lifted a ban on product placement (paying for the inclusion of brands within programming), directly addressing the problem of PVR-aided ad skipping. References to opportunities for 'AFP' (advertiser-funded programming) would quickly become standard features of broadcasters' commissioning briefs. In November 2011, the BAI introduced an increase (to twelve minutes per hour, twice the limit imposed on RTÉ) in the amount of advertising that commercial television stations could carry. Finally, in April 2012 the BAI began accepting applications from music-driven stations to derogate from the statutory requirement to devote at least 20 per cent of airtime to news and current-affairs content.

Though such measures clearly improved the viability of commercial broadcasters, they simply could not compensate for the scale of the revenue decline in the core advertising market. Such straitened circumstances amplified existing tensions between public and commercial media (print and broadcast) over territorial demarcations. An *Irish Times* editorial on 25 April 2009, while sympathizing with RTÉ's financial woes, nonetheless noted that:

> if advertising revenues are down, then those parts of its service which RTÉ itself defines as purely commercial should bear the brunt of any costs. The organization needs to be far more transparent about the demarcation lines between its commercial and non-commercial activities. It has to be much clearer about how its online strategy, exemplified by the impressive RTÉ player launched this week, dovetails with its public service remit.

This was echoed by a sustained campaign led by the National Newspapers of Ireland (NNI) from 2010 complaining that RTÉ was using publicly funded content to compete with commercial media in the Irish online-advertising market. By 2010, rte.ie could reasonably claim to be the most popular Irish media website, attracting 3.7 million unique monthly users (RTÉ Annual Report, 2010: 28), as compared with the 3.2 million visiting the independent.ie (INM Annual Report, 2010: 7). Thus, at a point when the *Irish Times* and the *Irish Independent* were trying to convince smartphone owners to pay for their branded news apps, NNI asserted that RTÉ was unfairly competing by offering their – publicly subsidized – content via their app for free.

RTÉ suggested that it would be 'perverse' if licence payers were prohibited from accessing licence-fee-funded content online (Flynn, 2014). In any case, given that RTÉ's online revenues had only accounted for €2.5 million or 3 per cent of the €97 million Irish online-advertising market in 2009, the station suggested that NNI's real competition came from the new media giants increasingly establishing their European headquarters in Dublin's docklands (*IT*, 16 July 2009).

Commercial and public broadcasters also tussled over the Sound and Vision fund. Noel Curran queried why broadcasters like TV3 and Setanta supported by 'hugely capitalized' owners should have access to public funding. For its part, in June 2009, TV3 complained that it was being short-changed by the Sound and Vision Fund: it had received just €5 million of the €45 million dispensed for television projects since 2003, as compared to the €18 million that went to TG4 and the €15 million that went to RTÉ.

TV3's David McRedmond would later complain that the Sound and Vision fund should better reflect the market share of those stations that applied to it, framing it as a reward for ratings success, even if the latter were driven by highly commercial simulcast content. RTÉ none-too-subtly hinted that this was already the case, pointing out that TV3 had won funding for 'highly populist programming' despite the Sound and Vision scheme's remit to support 'high value projects of a cultural bent' (*IT*, 9 July 2009).

In this regard, 2013 saw the BAI propose a reorientation of broadcast funding that might have mollified both the public and commercial sectors' concerns. The proposal grew out of what was a key motivation for introducing the 2009 Broadcasting Act in the first place: the need to assure the European Commission that state support for public-service broadcasters was transparent, and proportional to overtly defined public-service objectives for RTÉ and TG4. In addition to offering this definition, the 2009 act provided for the appointment of an independent body – the BAI – to conduct both small-scale annual reviews of the performance of the public-service broadcasters, but also much more in-depth assessments of their long-term strategies at five-year intervals. Both sets of reviews would inform government decisions on the level of public funding made available to such broadcasters.

The outcome of the first review in the summer of 2013 proved interesting. Although RTÉ had submitted a range of future-scoping exercises across a low-cost to high-cost continuum, the consultant appointed to assess them concentrated on RTÉ's minimalist 'base case' scenario, arguing that costlier options lacked sufficient detail (Crowe Horwath, 2013). On that basis, the consultant concluded that RTÉ required no additional public funds. TG4's submission was far more ambitious, but the €11 million in new public funds sought to achieve this was considered politically impossible (TG4's s total budget at the time was €36 million). The BAI broadly accepted the Crowe Horwath conclusions with regard to TG4's requested funding increase (which was politely – but indefinitely – deferred), but adopted a more nuanced response with regard to RTÉ.

The BAI endorsed the consultant's recommendation that RTÉ should explore further potential for cost reduction (not least by considering more extensive use of independent production) and seek greater commercial revenues. However, the regulator also argued that the base case put forward by RTÉ was inadequate to its mission to provide 'attractive and culturally relevant services to the Irish audience' (BAI, 2013) and, contra to the consultant's advice, recommended that a greater level of public funding be made available to RTÉ (without specifying how much). Crucially, this was to be accompanied by a rebalancing of RTÉ funding away from its reliance on commercial revenues:

> The Minister [for Communications] should determine a point above the current level of licence revenue where further public funding will be matched by a reduction in commercial revenue for RTÉ achieved through a restriction on commercial activities by, for example, a reduction in permitted advertising minutage, or a reduction in sponsorship or a combination of both. (BAI, 2013)

If accepted such measures might have assuaged the concerns of both commercial and publicly owned broadcasters. In the event, however, the BAI's recommendations were effectively ignored by successive ministers for communication and the status quo, equally unacceptable to all parties, was left undisturbed. Noel Curran could reasonably observe in an April 2016 speech that RTÉ – then preparing to enter into a new five-year planning cycle – was doing so 'without any meaningful response to the first one' (*IT*, 14 Apr. 2016).

The political reluctance to support a licence-fee increase stemmed from a combination of factors, not least the ongoing generalized suspicion of anyone who routinely asked questions about the political process. Given RTÉ's responsibility to reflect public protest at the austerity-era policy measures, there were ample opportunities for politicians of all stripes to be affronted at perceived media bias. In March 2012, having suggested that RTÉ was encouraging people to break the law by giving airtime to anti-property-tax campaigners, Fine Gael's Leo Varadkar asserted that RTÉ was biased towards centre-left/liberal parties (*IT*, 13 Mar. 2012). Yet in March 2015 a senior figure within the centre-left, Labour's Pat Rabbitte, would instead describe RTÉ as a 'recruitment sergeant' for the far left and Sinn Féin.

These were hardly auspicious conditions for seeking approval for increased public funding. Political mistrust was also now accompanied by a wider suspicion of broadcasters: Eurobarometer surveys carried out between November 2009 and November 2015 saw the percentage of Irish audiences who trusted television fall from 68 per cent to 55 per cent. Radio suffered less, but its decline from 71 per cent to 64 per cent over the same period was hardly encouraging. This followed a series of – at best questionable and at worst, just plain wrong – television news and current-affairs editorial decisions that appeared motivated less by abstract considerations of the public interest and more by crude pursuit of ratings.

An early example was TV3's decision to lead their St Stephen's Day 2009 news bulletin with a story revealing that the minister for finance, Brian Lenihan, had oesophageal cancer (Rafter and Knowlton, 2013). Although this did not breach regulatory codes, the public response was almost universally negative. The insensitive timing rankled most: Lenihan had only been aware of his diagnosis for some weeks and, as of Christmas 2009, had not yet revealed his condition to all of his close friends and family. TV3's alert to Lenihan that they were going to run the story effectively gave him just forty-eight hours to let those close to him know of his condition (to which he would sadly succumb eighteen months later).

However, it was RTÉ that inflicted the most spectacular own goals in this period, via two 2011 broadcasts: a *Prime Time Investigates* edition called 'Mission to Prey', and a live debate between candidates for the presidential election on the Pat Kenny-hosted *Frontline.*

Broadcast in May 2011, 'Mission to Prey' recounted the story of how, in 1982, Father Kevin Reynolds, an Irish priest working in Kenya had not only raped a teenage girl but, after she became pregnant, prevailed upon her not to reveal his involvement. One sequence showed the priest – now back in Ireland – being confronted by an RTÉ reporter with the allegations as he emerged from officiating over a first-communion ceremony. Despite the manner of this approach, Father Reynolds assented to an interview in which he denied the allegations. He would continue to do so via a series of legal letters to RTÉ – to the extent of offering to take a paternity test – in the fortnight between his interview and the broadcast.

The broadcast provoked huge public reaction: Father Reynolds was asked to step down from his ministry, and the allegations were referred to the Garda's sexual-assault unit. In September 2011, in the course of a high-court case for defamation brought by Father Reynolds, the priest submitted a paternity test proving he was not the father of the by now adult woman. In October, five months after the original broadcast, *Prime Time Investigates* issued a lengthy apology to Father Reynolds, acknowledging that the programme should never have been broadcast. A month later, Noel Curran described the broadcast as 'one of the gravest editorial mistakes' ever to occur in RTÉ (*IT*, 23 Nov. 2011).

By that stage, though, there was another contender for that category. In October 2011, in the final televised candidate debate of the 2011 presidential election campaign, *Frontline* host Pat Kenny read out a tweet purporting to originate from Sinn Féin, implying that the leading candidate, businessman Seán Gallagher, was not as politically independent as he claimed to be. The tweet (and Gallagher's responses to it) destroyed his poll lead and three days later Labour's Michael D. Higgins emerged as the victorious candidate. An election-day opinion poll found that 35 per cent of voters were influenced by the *Frontline* broadcast.

It subsequently emerged the tweet had not come from Sinn Féin, but rather from a Twitter account carefully constructed to resemble that of the official Sinn Féin presidential campaign, thus raising doubts as to the accuracy of its content. Members of the *Frontline* team had been made aware of this by Sinn Féin some 20 minutes before the debate concluded but, apparently not recognizing its significance, they failed bring it to the attention of the candidates or the audience. Again it took four months for RTÉ to formally apologize to Gallagher.

Of the two incidents, 'Mission to Prey' was unquestionably the more damaging. It literally cost RTÉ dearly: Father Reynolds' defamation case concluded with an approximately €1 million settlement and, using new powers under the 2009 Broadcasting Act, the BAI levied a €200,000 fine on RTÉ for breaching fairness and privacy codes. There were personnel costs too. All of the key figures directly associated with the programme were moved out of current affairs and two – reporter Aoife Kavanagh and Head of News Ed Mulhall – left RTÉ entirely. Noel Curran offered his resignation but was rebuffed by the RTÉ board on the grounds that a decapitated head learned no lessons.

Questions persisted regarding an internal RTÉ culture that dismissed Father Reynolds' willingness to take a paternity test as a mere delaying tactic. The answers emerged over a series of investigations: two internal to RTÉ (one conducted by Press Ombudsman John Horgan, and the second by former Northern Ireland Ombudsman Maurice Hayes), and a third commissioned by the BAI's compliance committee, conducted by Anna Carragher, a former controller of BBC Northern Ireland. At a prosaic level, the various reports pointed to questionable day-to-day journalistic practices (failure to seek sufficient corroboration of sources, poor note-keeping); the absence of any guidelines on the use of door-stepping; and a failure to adequately involve RTÉ's legal department in deliberations as to how to respond to Father Reynolds' denials. In a similar vein, the *Frontline* incident revealed a lack of formal protocols within RTÉ for dealing with social-media contributions: although new media could potentially broaden the voices brought into the public sphere, RTÉ had not yet learnt how to guarantee their authenticity.

'Mission to Prey' also drew attention to broader issues including a degree of groupthink within the *Prime Time Investigates* team and, by extension, RTÉ as a whole. By 2011, it could be assumed, decades of church-related scandals made many willing to countenance the allegations about Father Reynolds, but this could not excuse the station's failure to verify them. Furthermore, both 'Mission to Prey' and the *Frontline* broadcast drew the accusation that the station's approach to current affairs, by prioritizing drama over accuracy, was adopting an increasingly tabloid approach motivated by audience chasing. This was reflected in subject choices and the increasing recourse to door-stepping as a means of generating on-screen tension. One former RTÉ journalist wrote of *Prime Time Investigates* that:

> In more recent times the programme appears to have become increasingly concerned with the peccadilloes of petty criminals and the misdeeds of the marginalised … recasting … the *Prime Time Investigates* strand in a more populist guise, with greater elements of drama and tension, telling more stories that were less analytical but that yielded to a simply black and white treatment in which the baddies got their comeuppance in the end. (*IT*, 12 May 2012)

Horgan's report referred to the risks associated with investigative journalism. Of its nature, investigative journalism is speculative, and thus potentially unproductive. Though tolerable in the context of healthy revenues, the financial strictures facing RTÉ in 2011 made it harder to consign months of work to the wastebasket, creating pressure to broadcast material that might in other circumstances have been treated with more caution.

Acknowledging this, Curran committed to a new approach to the scheduling of current-affairs output:

> They will be broadcast when they are ready, not when everyone is working towards a date in the schedule that they have been given three months in advance. The scheduling of back-to-back episodes was a mistake because it puts too much pressure on production teams to handle so much controversial matter at once. (*Drivetime*, RTÉ Radio 1, 3 Apr. 2012)

In March 2012, after the BAI upheld a complaint by Seán Gallagher regarding RTÉ's failure to publicly identify the source of the *Frontline* tweet, Curran committed to update the station's social-media guidelines. Two months later, RTÉ sent the minister for communications a report committing to fully implement the recommendations of the various 'Mission to Prey' reports. There were new staff guidelines, curbs on surreptitious filming and doorstep interviews, earlier involvement of legal advisers and an examination of scheduling pressures and of the broader culture in which RTÉ decision-making occurred. In particular, staff involved in major investigative programmes would henceforth be required to submit proposals for programmes to a senior editorial board (potentially including the DG) in advance of broadcast.

The high-profile appointment in July 2012 of Kevin Bakhurst, former controller of the BBC News Channel, as RTÉ's new managing director of news and current affairs would help rebuild the reputation of the *Prime Time* investigation unit, which, after a period of suspension, went on to produce a body of compelling work. Despite RTÉ's earlier sterling exposé of institutional child abuse, 'Mission to Prey' undermined not just RTÉ's credibility but also its moral authority to subject other institutions to criticism. The financial cost associated with the 'Mission to Prey' defamation suit and that potentially resulting from a Seán Gallagher suit launched in 2013 also created a climate of caution that sometimes encouraged discretion within the station rather than valour. In January 2014, Rory O'Neill, otherwise known as the drag queen Panti Bliss, went on *The Saturday Night Show* to discuss growing up gay in Ireland. Pressed by presenter Brendan O'Connor to cite examples of homophobia, O'Neill mentioned two *Irish Times* columnists – John Waters and Breda O'Brien – and members of the Catholic conservative lobby group, the Iona Institute. The two named individuals plus four Iona members complained that they had been defamed. By contrast to the station's delayed response to the Father Reynolds and Gallagher affairs, and without consulting O'Neill, RTÉ immediately apologized and – on in-house legal advice – paid out €85,000 to the complainants. Ironically, although the broadcast generated more than a thousand protests to RTÉ and the BAI, the bulk of these related not to O'Neill's comments but to the decision to make the payout, which was framed as funded by the licence-fee payers.

If television was experiencing travails in terms of content and ad revenues the much healthier state of another part of the sector went relatively unnoticed. By 2012, total pay-TV revenues of €650 million (mainly accruing to Sky and UPC) far outstripped the €200 million television advertising market. Although commercial broadcasters railed against RTÉ's market dominance, Sky alone accounted for 43 per cent of all Irish television revenues in 2011, its €382 million turnover outstripping RTÉ's by €32 million. UPC had attempted to match Sky's television-subscription success, but found itself outgunned by Sky for content. In July 2010, the satellite giant announced the launch of Sky Atlantic, built around an exclusive deal with the US channel HBO for access to its drama output. In 2014, the unthinkable occurred when Sky Sports acquired rights to fourteen out of forty-five championship games sold by the GAA that year. Domestic audiences for such games collapsed: fewer than 35,000 typically watched the GAA games on Sky, as compared with the 250,000-plus previously secured by RTÉ and TV3. Overall, however, such content saw Sky's subscribers surge from 500,000 in 2007 to nearly 700,000 by early 2012, while UPC's fell to less than 500,000 by the close of 2010. However, UPC's business model increasingly emphasized broadband download speed over channel choice as it became possible to deliver the long-promised 'triple-play' package of television, broadband and home phone to a wider range of customers. As peak speeds increased so did broadband subscribers, more than doubling (to 215,000) between the beginning of 2009 and mid-2011. In effect, UPC increasingly operated as a telecommunications-style 'common carrier', whereby television channels were just one instance of data traversing its network.

This logic of convergence worked two ways: in July 2011, Eircom announced it was entering the pay-TV market. Faced with such competition (Eircom was not just a triple-play but a quad-play operator), in February 2013, Sky signed a deal to lease space on British Telecom's fibre network, adding broadband and home phone lines to its television package. By September 2015, a month in which Eircom was rebranded as Eir and UPC as Virgin

Media, the broadband market was split between Eir on 35 per cent, Virgin Media on 29 per cent, Sky on 9 per cent and Vodafone on 18 per cent.

Despite the development of the private platforms, some 250,000 Irish television households still lacked a digital television service. This had become a critical issue by 2010, as the EU-mandated analogue signal switch-off date of 2012 neared. RTÉ had already commenced this process in radio, announcing the end of medium-wave transmission of RTÉ1 in January 2008 and, in October 2014, the closure of the long-wave radio service. Both decisions were legitimated by reference to the low use and high cost of the services, but both were criticized by, in particular, emigrant groups who argued that they constituted a lifeline for the Irish overseas. RTÉ's response simply emphasized their investment in digital alternatives, the geographic reach of which extended far beyond the analogue radio services.

Government vacillation regarding the precise structure of a digital terres-trial television service notwithstanding, RTÉ had continued working through 2008 to complete the €70 million underlying national DTT infrastructure. As trials neared completion, it appeared that a commercial DTT service to rival the cable and satellite platforms was finally in the offing. In August 2008, after a competition involving three consortia, the BCI awarded Boxer, a Swedish company with Communicorp backing, a contract to operate three commercial DTT multiplexes. Boxer's application had been bullish about its prospects, promising to spend €145 million rolling out the service and expecting to secure 215,000 subscribers and €50 million in revenues by 2013. However, the recession intervened. Though in July 2008, Boxer (and Communicorp) Chairwoman Lucy Gaffney had suggested that the key role of television in everyday life made it almost recession proof, in April 2009 Boxer withdrew from its commitment to offer a DTT service, citing 'prevailing and anticipated economic circumstances' (*IT*, 21 Apr. 2019). The BCI reverted to the second-placed bidder, One Vision, but it too demurred on broadly similar grounds. In August 2010, the BAI announced that commercial DTT was on the back burner, at least until the switch-off of analogue services in 2012.

This effectively left DTT in the sole possession of RTÉ Networks Limited. A full-scale national trial of the service, now named 'Saorview', commenced in October 2010. RTÉ promised that the new service would raise the percentage of the population able to access domestic channels from 95 per cent under the analogue system to 98 per cent under Saorview. Furthermore, it was suggested that a parallel service using satellite (Saorsat) might not only reach the remaining 2 per cent but also bring high-speed internet access to the most remote regions of the country. The following March (2011) saw a public-awareness campaign for the digital switchover launched to promote the eight local channels available via the service.

The October 2012 digital switchover occurred in somewhat muted circumstances. There was a sense that DTT was very much the poor relation in terms of delivering digital television. Mindful that its core demographic – the 180,000 homes for whom it was the sole means of accessing digital television – was weighted towards older, lower-income families in rural areas, TV3 had initially expressed reservations about paying to have its channels included on the platform. Optimism regarding the prospects of adding a commercial DTT service down the line was somewhat undermined by the blunt conclusions of a 2013 BAI-commissioned report in 2013:

> Investors are likely to judge DTT a less appealing technology to fund than it was five years ago when fully 30 per cent of Irish households were still watching analogue TV and digital switchover was a driver of migration to digital platforms; the time horizon for investments to reach profitability has also shrunk, and projects which were marginal in 2008 will not be invested in today's economic climate. (Oliver and Ohlbaum, 2013: 5)

Indeed the report expressed doubts about the viability of the existing Saorview service as intensified price competition between UPC, Sky and Eir created new incentives for up to half of Saorview's core market – the 180,000 homes exclusively served by DTT – to move to other television-distribution systems. The bald conclusion then was not merely that 'the BAI should not therefore proceed with a new licensing process' (48), but that the existing Saorview service should be replaced by mandating those broadcasters already on the service to broadcast unencrypted signals via the Astra digital satellite.

Although the BAI abandoned the notion of a new licensing process, these pessimistic predictions proved incorrect. By March 2015, Saorview had added 100,000 customers to the 500,000 using the service at its 2011 launch, and more subscribers joined thereafter. By one measure Saorview was the largest platform in Ireland by January 2016, with 676,000 homes connected, even if only 186,000 were Saorview only. Saorview's success probably owed much to the widespread availability of broadband, which facilitated the practice of 'cord-cutting' or 'cord-shaving', whereby users abandoned television subscription packages via satellite/cable and relied instead on content streamed or downloaded online. In the 1970s, cultural theorist Raymond Williams had used the 'flow' metaphor as a way of characterizing the manner in which channels constructed – and audiences consumed – an evening's worth of entertainment. By the twenty-first century, the 'file' metaphor appeared to better capture the manner in which viewers, especially in younger households, became more used to 'curating' their own viewing schedules through reliance on PVRs, on-demand services or the less-than-legal avenues afforded by high-speed downloading of content through torrent sites. In that context, the one-off payment required for Saorview took care of the domestic-content element of such content jigsaws.

The prevalence of such viewing patterns was amply demonstrated by the Irish launch of the Netflix on-demand service in January 2012. Despite a limited range of content relative to the US version (due to rights issues), the €6.99 monthly charge succeeded in attracting an estimated 150,000 Irish subscribers by August 2013. What made this all the more remarkable was that until September 2016, when it appeared on Virgin Media's cable package, Netflix had to be watched via a computer, tablet, smartphone or games console. With an estimated 20 per cent of Irish homes subscribing to the service by the end of 2016, Netflix's spread reflected the emergence of completely novel screen-viewing patterns, for which a television set was optional. It also prompted responses from existing players: UPC and Sky both introduced their own on-demand television-and-movies service – in May and September 2012, respectively. Even RTÉ, in an 'if you can't beat them, join them' September 2012 deal with Netflix, made it possible to access 140 episodes of RTÉ drama and comedy content via the platform.

Irish television had already adapted to on-demand services long before Netflix. Following the success of Channel 4's 4OD and the BBC's iPlayer online 'catch-up' services from 2006, in April 2009 RTÉ launched the RTÉ Player, permitting on-demand viewing of shows for twenty-one days after initial airdate. (TV3 would launch its own player in October 2011.) Though initially exclusively available via PC, from 2010 the player began to migrate to other intermediary technologies, such as games consoles. The player conjured audiences for shows that might not have justified an existence based on their broadcast ratings. *Fade Street*, an RTÉ reality show, won only a modest live broadcast audience (100,000 viewers) from November 2010, but added 69,000 views on the player in the ten days after broadcast. By August 2013, although one survey suggested that 91 per cent of viewers were still watching television on a television set, a sizeable minority – 33 per cent – were also using portable tech like laptops, tablets and smartphones to do so (Eir, 2014b: 18). A particularly attractive feature of the player for broadcasters, especially in an era of PVR-based viewing, was that audiences had to tolerate up to ninety impossible-to-skip-through seconds worth of advertising before accessing the sought-for content. Furthermore, the player offered an opportunity to associate views with user profiles, emulating the ability of social media sites to offer targeted audiences to advertisers. From May 2013, TV3 required users of its player to register before viewing: though the player was nominally free, data about the registered users constituted a commodity in its own right.

As the popularity of on-demand viewing developed – by 2014, the RTÉ Player app had been downloaded over 1 million times, and the RTÉ Jnr app aimed at children was securing 1 million views per month across its online platforms – the players came to be regarded less as ancillary services and more as stand-alone platforms with their own content. By August 2014, RTÉ Head of Television Glen Killane was including digital-audience targets in programme planning, reflecting a 'digital-first' strategy with content addressing younger audiences – especially comedy – increasingly appearing on the player before its 'linear' broadcast.

The prevalence of consumption via such digital platforms repeatedly drew attention back to an issue first raised by Eamon Ryan towards the end of his 2007–11 tenure as minister for communications: how to deal with those who, because they did not own a television, did not pay a television-licence fee, but nonetheless availed of public-service content via online platforms. Other European countries had already begun to grapple with this question in a variety of ways. In 2006, the Danish parliament agreed to replace the television licence with a media licence requiring anyone with either a television or internet-capable device to pay a licence fee. In Finland and Sweden, the 2010s saw the flat-rate licence replaced by a progressive broadcasting tax paid by all households regardless of whether they owned a receiver of any kind or not, reflecting the view that, whether one used public-service content or not, it was socially important and thus worthy of public support.

In Ireland, policy emerged in a less coherent fashion. Although in May 2008 Ryan stated that, in order to encourage innovation in the development of digital applications, those who watched television via mobile phones or PCs would be exempt from the a licence fee, within two years he would argue for a new revenue stream for public-service broadcasting reflecting the emergence of new screen technologies. By January 2012, his successor, Pat

Rabbitte, would introduce a time frame for replacing the television licence with a 'household broadcasting charge', and from December 2013 onwards, 1 January 2015 was routinely cited as the launch date for the new system.

The appeal of the system was obvious. Though the government was reluctant to be seen to be granting RTÉ an indirect funding increase, the household broadcasting charge promised to increase by €20 million to €30 million the amount of public funding raised for broadcasting. It could achieve this without increasing the amount levied because, in theory, it promised to reduce the 10–15 per cent licence-fee-evasion rate to zero. RTÉ, who had long complained of An Post's inefficiency in collecting the licence fee, clearly favoured the new charge. More surprising was the fact that so too did elements of the commercial media. In December 2011, Tim Collins, chief executive of Ocean FM (and later of Newstalk), tacitly invoked the argument that the recessionary collapse of the 'advertising for journalism' quid pro quo raised the need for new support for public-service broadcasting. However, Collins opined that much of this was already present on independent commercial stations. Months later, in February 2012, Thomas Crosbie Holdings' Alan Crosbie made the logic of Collins' reasoning explicit, arguing that commercial media should receive a share of the new household broadcasting fund on the grounds that they performed a public service and that advertising revenues were 'fickle'.

The state response to this was cautious. At the April 2012 Independent Broadcasters of Ireland conference, minister Pat Rabbitte reminded delegates that commercial broadcasters had sought and accepted licences on terms that included an acceptance of commercial risk. In the same month, Rabbitte appeared to rule out redistributing household-broadcasting charge revenues to newspapers because, unlike broadcasters, they were not subject to a set of statutory obligations. The commercial sector found itself confronted with the same thorny issue it had laid at RTÉ's door since the 1990s: how to definitively and objectively distinguish public-service content from purely commercial fare? Calls from commercial media for such funding were rarely accompanied by detailed attempts to address this question. In any case, as Rabbitte pointed out in April 2013, by which stage 'pretty much everybody' had sought additional finance from the mooted fund, its limited size could hardly address all of the problems facing Irish media.

By the beginning of 2013, there was at least a rhetorical sense that the recessionary spiral had bottomed out. Although spending on public services remained well below its 2008 levels, unemployment rates which had peaked at

15.2 per cent in 2012, persistently fell thereafter. The crisis-induced levels of economic emigration peaked in 2013, and though Ireland still faced a crippling debt to the IMF, EU and European Central Bank, it was at least able to exit the bailout programme that had severely limited the state's economic sovereignty since 2010.

This economic improvement was reflected in the Irish advertising market. Having fallen consistently since 2007, advertising expenditure across all media plateaued in 2013 and even increased slightly within individual sectors. However, online-advertising growth rates continued to far exceed those in other media. By 2013, figures from WARC, the international advertising research company, suggested that online revenues accounted for 23 per cent of all Irish ad-spend, second only to the 28 per cent accounted for by television. By 2015, online had become the single largest destination for Irish advertising expenditure, accounting for 40 per cent of the total. Only a small proportion of this remained in Ireland, however. IAB Ireland estimated that Google alone accounted for 52 per cent of the €445 million spent in the Irish online-advertising market in 2016, with Facebook accounting for a further 14 per cent.

Nor did the post-2013 stabilization of overall television ad revenues (at between €225 million and €250 million) necessarily benefit indigenous stations. By 2013, somewhere in the region of €50 million went to UK-based channels that made little or no investment in Irish production. In 2006, UK advertising opt-outs had accounted for 12 per cent of all Irish commercial impacts (a measure of how often an ad was viewed), but this grew to 23.5 per cent by 2013 as the number of opt-out channels grew to thirty-six. These were further augmented by Channel 4's November 2013 announcement that it would begin direct sales to the Republic of Ireland from the following January.

Such figures clearly shaped the thinking of legacy media as they struggled with the new dispensation. In January 2013, Noel Curran held a series of meetings with RTÉ staff at which he asserted that 'the worst of the cuts are over'. It was certainly hard to imagine where additional savings might be made after a 2012 in which another voluntary redundancy scheme had sought 200–300 more job cuts, the sports budget had been cut by 25 per cent and 10 per cent had been cut from overseas acquisitions. RTÉ's London office had closed in September 2012, and regional studios in Waterford, Athlone, Sligo and Dundalk were only saved by relocating RTÉ offices to institutes of technology in those towns and cities.

RTÉ's key financials improved in 2013 and 2014. It recorded modest surpluses (the first since 2008) and even permitted new staff hires by 2015. Yet its largely static revenues limited opportunities for significant new investments in programming. For a period, the impact of this on drama was disguised by the remarkable success of *Love/Hate*, a Dublin-set crime drama, which, having won healthy audiences from its October 2010 debut, had become something of a phenomenon by 2013. Some 970,000 viewers tuned into the fourth season opener and, by the time of the season-five finale in autumn 2014, average audiences surpassed 1 million per episode. The show generated significant earnings for the station through direct-to-consumer sales of box sets and sales to international broadcasters. By March 2015 the show had been sold in thirty foreign markets.

However, when *Love/Hate* went on an extended hiatus after 2014, the drama cupboard looked a little bare. RTÉ would screen just eight hours of new Irish drama in 2015, and when channel controller Bill Malone had looked to rebrand RTÉ 2 Television as a youth-oriented (under-35) channel from autumn 2013, the tools in his arsenal were less drama and more an emphasis on what was politely described as 'light factual' programming (celebrity-led investigative shows) and, by late 2014, sketch-based comedy.

On radio, strategies to retain, if not necessarily build, audiences relied on high-profile presenters. Yet, in August 2013, Pat Kenny, arguably RTÉ's biggest star, announced that he was moving to Newstalk. Kenny's unhappiness with his position in the RTÉ hierarchy was an open secret. Originally devised as a vehicle to allow him to return to a post-*The Late Late Show* current-affairs slot, *Frontline* had effectively been downgraded to an element within the prime-time schedule after the Sean Gallagher tweet incident. The loss was hugely symbolic for RTÉ – Kenny had been a stalwart at the station for forty-one years – but also a potential boon for Newstalk, which spent €1 million in an attempt to convince Kenny's fans to follow him to the new station. Some clearly did: by summer 2014, Kenny had nearly tripled Newstalk's mid-morning audience, although his 134,000 listeners still trailed the 295,000 listeners that his replacement at RTÉ – Sean O'Rourke – managed to retain. In a counter-strike eighteen months later, RTÉ would poach Today FM ratings leader Ray D'Arcy, putting him into an afternoon-radio slot and a weekly Saturday-night television chat show.

Investing in content might raise revenues, but RTÉ needed more secure sources of funding. An RTÉ-commissioned report by PWC, published in April 2013, emphasized the imbalance in the station's finances: while the €160 licence fee fell far short of the European average (€216), RTÉ's proportional reliance on advertising was twice that of the European average for public-service broadcasters. The state's response was to further reduce the station's licence-fee income after the minister for social protection, Joan Burton, announced in October 2013, that her department would cut €5 million from its annual payment to RTÉ for pensioner licences. Under these circumstances, RTÉ began to reconsider existing carriage deals with satellite and cable operators. As a 'must-carry' station under the 2009 Broadcasting Act, Sky and UPC were obliged to include RTÉ channels on their platforms; both received RTÉ at no cost. In 2014, RTÉ commissioned Mediatique to assess the value of transmission fees. The research concluded that although RTÉ would lose substantial ad revenues if it were removed from Sky and UPC, the platforms stood to lose far more – €19 million per annum for Sky and €11 million for UPC – from cancelled subscriptions. By the end of 2014, therefore, RTÉ began to discreetly lobby for some recompense from platform operators for their content.

However, for Noel Curran, the long-term goal for RTÉ was to colonize more of the online market. He had been complaining of RTÉ's failure to make more inroads in this regard since his appointment, and in 2011 had committed to increasing the proportion of RTÉ's digital revenues from 2.5 per cent to 15 per cent. Four year later, in March 2015, although half the population was using RTÉ's digital services in any given week, digital revenues still accounted for only 9 per cent of RTÉ's total commercial revenues. Despite the manifest popularity of some digital content – online only content drove the number of monthly streams via the RTÉ Player up to 4.4 million in April 2016, from 3.4 million in 2015 – these audiences proved hard to monetize. Nonetheless, RTÉ's digital strategy appeared to focus on developing streaming services such as the RTÉ Player. In May 2014 RTÉ had announced GAAGO, a partnership with GAA to stream the hurling and football championships to anywhere with an internet connection – for an annual subscription of €110. In February 2015, RTÉ had announced plans to make its player available outside Ireland for a fee of €8.99 a month, offering 400 hours of content built around recent popular domestic shows.

TV3 seemed less inhibited in its attempts to exploit content online. A month after an October 2012 deal making its content available to US and Canadian audiences via the Roku streaming service, the station announced plans to introduce micro-payment systems, charging viewers 49p per view to watch content via the 3Player after an initial seven-day free period elapsed.

This followed a December 2012 trial whereby 6,000 people had paid €1.49 to watch pre-broadcast episodes of the channel's scripted reality show *Tallafornia*. The channel also played with hybridizing content. By March 2013, the station's *Xpose* entertainment news show had become a multi-revenue-stream production, as a print magazine tie-in launched in December 2012 was complemented by a digital version accessed via mobile app available for €1.79 per issue or €13.99 for an annual subscription.

The sums involved in such transactions seemed trivial alongside the potentially seismic shift resulting from UTV's November 2013 announcement of UTV Ireland, a new Dublin-based station which would provide the ITV schedule to the Republic of Ireland with content customized to the preferences of local viewers. The new channel would have a hundred staff and a particular focus on news and current affairs. Betting that a sustained recovery was around the corner, UTV hoped to grow their 2.9 per cent of the Southern Irish market to a point where the new channel would surpass TV3's 11.4 per cent share. Securing exclusive Irish rights to ITV soaps *Coronation Street* and *Emmerdale* was critical to this strategy. Though simulcast on UTV and TV3 to Irish viewers, they routinely secured far larger audiences in TV3 – the latter's highest weekday audiences. In November 2013, UTV announced a deal with ITV depriving TV3 of both soaps from early 2015.

TV3 moved swiftly to mitigate the impact. A contractual renewal with Fremantle Media secured some ITV 'shiny-floor' shows (including *X Factor* and *Britain's Got Talent*), while the costume-drama ratings juggernaut that was *Downton Abbey* was retained following a for-the-life-of-the-show deal with NBC/Universal. But there was also a particular emphasis on increasing the volume of local production. By June 2015, six out of ten of TV3's top shows would be Irish-made, up from just one of ten in 2014. These included the existing stable of localized reality formats – to which rights for *Gogglebox* were added in May 2014 – but also new internally generated game-show formats such as *Algorithm*, and *The Lie*. Taken together, such shows suggested an increasing emphasis on programme sales as a means of generating revenue, reducing dependence on the crowded advertising market: forty-eight channels sold advertising in Ireland by mid-2014. However, TV3's most direct response to UTV Ireland was to proceed with the – already commenced – commissioning of a locally produced soap. By February 2014, a contract was awarded to the Irish production company Element Pictures in association with a UK producer, Company Pictures. Substantially supported by the BAI's Sound and Vision Fund (which had granted over €2 million to the show by the end of

2016), the production saw €7 million and over 100 crew and talent expended to produce 104 twice-weekly twenty-two-minute episodes for the first season. Set around a police station in a Dublin harbour town, *Red Rock* debuted in January 2015. Critically well received, the show garnered audiences of 250,000-plus (including catch-ups via the TV3 player) in its opening weeks, which, though well below those attracted by UK soaps, still occasionally breached the daily top-ten of most-watched shows.

The competition between TV3 and the new UTV channel would prove relatively short-lived. The BAI had licensed the new station under Section 71 of the 2009 Broadcasting Act in February 2014. This meant a lighter touch from the regulator, but also no guarantee of access to the various television platforms. Nonetheless, in December 2014, the then minister for communications, Alex White, designated the channel as 'must-carry' for Saorview, accepting UTV Ireland's self-description as having a public-service 'character'. The fact that this character was largely based on a combination of ITV soaps and a nightly news and current-affairs hour raised doubts as to the new channel's prospects. The absence of any new local drama left UTV Ireland looking like an ITV opt-out service, albeit one lacking some of ITV's most popular shows, which remained on TV3.

Launched in January 2015, even UTV Ireland's high-profile news shows struggled to win more than 4 per cent of the audience. Though the soaps would help bring the channel's overall market share for January 2015 to 5.5 per cent, UTV revised its initial estimates for a year-one loss of €2–3 million up to €8.4 million by March, and again up to €16 million by June. (Even this proved optimistic – the real figure ended up at €19.5 million.) Remarkably, having spent more than a year developing UTV Ireland, by August 2015 UTV Media confirmed that it was in talks to find a buyer for both UTV and UTV Ireland.

There had been little in the way of media merger and acquisition activity in Ireland since 2008 – the collapse in advertising markets having made finding a buyer something of a challenge. The one prominent exception to this was Communicorp owner Denis O'Brien's building of a stake in Independent News and Media, which he'd been doing since 2006. O'Brien became the largest shareholder in April 2010 and reached 29.99 per cent in May 2012. Under stock-exchange rules, a 30 per cent or higher stake obliged the stockholder to make an offer to buy the entire company. His stake allowed O'Brien, who INM initially treated as a dissident shareholder, to reshape that board, adding three long-time associates in March 2009.

Simultaneously, after thirty-six years at the helm, Tony O'Reilly stood down as both a member of INM board and CEO. His replacement, his son Gavin O'Reilly, lasted three years before he found himself forced to resign in April 2012.

The situation whereby the owner of the largest (in term of audience-share) private radio group was also the single largest shareholder in the dominant force in Irish print media demanded some regulatory and political response. In May 2008, when O'Brien's holding exceeded 25 per cent, and again in March 2009 (after the appointment of new directors), the BCI reviewed Communicorp's radio licences to assess whether an 'undue amount' of communications media now lay in Mr O'Brien's hands. In the first review, the BCI expressed the view that Mr O'Brien's INM shareholding was insufficient to allow him to influence editorial content. In the second, having considered the totality of communications media (i.e. across all media) in Communicorp's franchise areas, the BCI concluded that since audiences could still access alternative perspectives from a range of other media, the businessman's print and radio shareholdings did not permit dominance in relation to opinion-forming power.

These decisions occurred against the background of the work of the advisory committee on media mergers, which had reviewed whether existing competition-law provisions could protect media pluralism and diversity in a June 2008 report to the minister for enterprise, Micheál Martin. Its conclusions echoed competition authority concerns about its own capacity to assess media diversity, but also drew attention to the lack of clarity regarding the criteria for assessing the impact of a media merger on diversity.

It recommended introducing both a statutory acknowledgement of the role of the media in maintaining a healthy democracy and a statutory definition of media plurality based on concrete indicators that took into account consideration of the implications of cross-media – including online – ownership. Finally, it emphasized the need to shift the regulation of media mergers away from a competition policy orientation, and towards a public-interest one, suggesting that since the question of media pluralism was 'one essentially of political judgment' (59), the primary locus of decision-making should lie with a politician rather than the competition authority. Given the degree of media attention attracted by Denis O'Brien's expanding INM shareholding, the committee report might have been expected to generate a prompt political response. However, six years would elapse before any of its recommendations found legislative expression.

Little progress was made under the Fianna Fáil/Green coalition up to the February 2011 election. In September 2011, seven months into his term as minister for communications, Pat Rabbitte had announced the preparation of legislation implementing the committee recommendations and providing for the collection of and publication of information on media ownership. The actual introduction of the legislation remained 'imminent' for three years, delayed in part by bundling it into a much broader Competition and Consumer Protection Act.

In the interim, 2012 saw the BAI publish a new ownership-and-control policy, which clarified some of the ambiguity in the relevant sections of the 2009 Broadcasting Act. The BAI policy distinguished between 'control of' and 'a substantial interest in' a radio station. The former referred to the power of a shareholder to 'determine' all aspects of company policy (including editorial output), whereas the latter acknowledged that even if a shareholding fell short of full control, a 'substantial' holding might allow a shareholder to 'influence directly or indirectly to an appreciable extent the strategic direction or policy (which shall include editorial policy)'.

Critically, however, the BAI could not arrive at a universally applicable definition of what Section 66 of the 2009 Broadcasting Act referred to as an 'undue' number of radio services or communications media, which might have served as an objective guide for setting limits to the extent of media-market control. The BAI concluded that there was 'no obvious practical matrix' for doing so, and determined instead to proceed on a case-by-case basis (BAI, 2012: 11).

Denis O'Brien's shareholding constituted an ongoing case as it reached 29.99 per cent in May 2012, prompting the BAI to write to Communicorp to ascertain the latter's 'position' with regard to INM. The situation was made more fraught by the conclusions emerging from the publication of the Moriarty tribunal report a year earlier. Moriarty had concluded that as O'Brien was bidding for the state's second mobile phone licence in 1995, the individual overseeing the process, the minister for communications, Michael Lowry, had received two payments and support for a loan from Mr O'Brien via intermediaries. In the careful language of the report, the 'reasonable inference' was 'that the motive for making the payment was connected with the public office of Minister for Transport, Energy and Communications' (Moriarty, 478). Lowry and O'Brien vigorously disputed the findings, denying any suggestion that such payments had been made or that there had been any corrupt contacts between the two men. Regardless, in July 2012 the

BAI argued that since, in their view, Denis O'Brien had a substantial interest rather than a controlling one, there was no requirement for it to rule on whether he held an undue amount of communications media.

If this seemed somewhat unsatisfactory, the 2014 Competition and Consumer Protection Act, when finally passed, was – at least notionally – understood to allow the state and its regulatory arms to address the high levels of concentration already characteristic of Irish media markets. Though still lacking quantifiable thresholds to prevent an increase in concentration of media ownership, the identification of a 20 per cent holding or voting strength as constituting a 'significant interest' suggested that any merger at or above that threshold would attract close scrutiny.

The 2014 legislation allowed the minister for communication to immediately approve media mergers, but also the discretion to refer them to the Broadcasting Authority of Ireland for a full investigation, regardless of the media types involved. Under the act, the minister could in turn appoint an ad-hoc committee to advise the BAI as to the likely impact of such mergers on pluralism and diversity. However, the act did not overtly permit retrospective application of media-ownership limits. Although comments from Pat Rabbitte during the drafting of the legislation appeared to anticipate that it might apply to existing instances of media concentration, his successor as minister when the legislation came into effect, Alex White, routinely stated that constitutional protections on property rights made retrospective application impossible. Though the 2016 report funded by the GUE/NGL European Parliament group, and commissioned by Sinn Féin MEP Lynn Boylan, pointed to legal opinion contradicting White's view, it became obvious that the government's will to challenge existing concentrations of ownership was in short supply.

The act also required the BAI to conduct three-yearly reviews of the impact on pluralism of media-ownership changes. The first of these, carried out on behalf of the BAI by Communication Chambers, concluded that ownership changes since 2012 had not impacted on media pluralism to any discernible degree, though with specific regard to Denis O'Brien this conclusion was based on the fact that Mr O'Brien's increase in ownership in INM to 29.9 per cent in that period did not 'seem to have created a significant interest, since such an interest likely existed already' (Kenny and Suiter, 2015: 6).

Given this, the real test of the act would lie in its operation going forward. And, as UTV's move to sell off UTV Ireland in summer 2015 signalled, in

the two years following the act's passage, there was a sudden flood of opportunities to trial it. In June 2015, it emerged that Liberty Global was
conducting due diligence on TV3. Having spent €265 million to acquire the
channel in 2006, the new reality of the Irish media market saw Doughty
Hanson looking for just €120 million for the two-channel group. In July 2015,
they accepted Liberty Global's even lower offer of €80 million. TV3 was a
relatively attractive prospect at this stage: it earned a €1.1 million operating
profit on €55.9 million turnover in 2013, and was attracting advertiser interest
in advance of the September 2015 Rugby World Cup, to which it had exclusive
Irish rights.

In October 2015, ITV acquired UTV and UTV Ireland for £100 million,
placing thirteen of the fifteen UK regional commercial licences in the hands
of a single company. (As the deal involved selling the 'UTV' brand, the
remaining radio holdings were renamed as the Wireless Group.) ITV's plan
to revive the Irish channel by reinstating the ITV content lost when it had
begun operations in January 2015 suggested a shift back to the opt-out model.
It was not clear how the substantial reduction in spending on UTV Ireland's
news and current-affairs output that this implied could be reconciled with
the public-service character on which UTV Ireland's Saorview placement
depended. In July 2016, apparently unable to square this circle, ITV sold
UTV Ireland on for €10 million to Liberty Global, which added it to the
TV3 stable. Job losses and a 'female-friendly' rebrand of the channel as 'Be3'
followed in December 2016.

The only remaining independent commercial Irish channel was Setanta
Sports, which had experienced something of a rollercoaster existence over the
previous decade. Having been valued at a reported €1bn in February 2008, the
loss of key English Premier League rights in February 2009 brought the
company to the verge of collapse. However, though forced to jettison most
of its international operation, an eleventh-hour investment by music
promoter Denis Desmond offered the original founders a lifeline. Though
operating on a smaller scale than before, the company maintained a niche as
a mixed free-to-air and subscription broadcaster based on a more modest set
of sports rights. These were substantially augmented by a three-year deal with
British Telecom (BT) in 2013. Having sought to move into the television
market, BT had acquired rights to the English Premier League and the
European Rugby Champions Cup to build a subscriber base in the UK. The
deal with Setanta made these available to an Irish audience.

Fittingly, having identified the same strategy as BT, by November 2015, Eir was in talks to acquire Setanta for €20 million to beef up its Eir Vision television platform (launched as eVision in October 2013). The deal was completed a month later and, by July 2016, the Setanta brand disappeared, replaced by 'Eir Sports'.

That rebranding exercise coincided with yet another acquisition. In July 2016, the Rupert Murdoch-led News UK offered £220 million for the Wireless Group's UK and Irish stations. The main attraction for News UK was the group's Talksport sports station and its exclusive Premiership radio rights. As such, the six Irish stations included in the deal (among which was FM104) were icing on the cake.

The final major acquisition move came in September 2016, when INM made a bid for the Celtic Media Group (CMG), a regional newspaper group based in the Midlands. Framed as essential to maintain CMG's viability, in November 2016 the Competition and Consumer Protection Commission approved the merger on the grounds that it would not lead to a substantial lessening of competition. This left the new minister for communications, Denis Naughten, who took on the role following the February 2016 election, to assess its impact on pluralism.

In all six cases, the acquisitions placed small and medium-sized media outlets into the hands of much larger corporate entities. Though former owners like Doughty Hanson, UTV Media and ITV were hardly minnows, the new owners were – with the exception of INM – not just multi-billion-euro international giants, but were simultaneously active across a number of media both within and beyond Ireland. In addition to being one of the largest global players in the cable-television market (and monopolizing that sector in Ireland), Liberty Global owned 'super-indie' producer All3Media, whose assets included *Red Rock* co-producer Company Pictures. In December 2015, Tony Hanway, the head of Liberty's Irish subsidiary Virgin Media, emphasized that Virgin Media cash would be used to both bolster TV3 bids for local sport content, but also to maintain production of exportable local drama and factual/game-show programmes. For its part, News UK's acquisition of the Wireless Group created a situation in Ireland whereby companies associated with Rupert Murdoch simultaneously occupied significant positions in the television-platform and broadcast-advertising markets through Sky Digital and Sky Television, the print market through the Irish *Sun* and *Sunday Times* Ireland, online (through its ownership of Storyful) and the radio market.

Yet for the most part these acquisitions failed, at least until late 2016, to attract close scrutiny under the provisions of the Competition and Consumer Protection Act, and were passed without a second-stage investigation. The public rationale for permitting them to proceed was captured in the boiler-plate language of the department of communications letters to the interested parties, which simply stated that the minister was 'of the view that the media merger proceeding as proposed will not be contrary to the public interest in protecting the plurality of media in the state'. The processes engaged in, and conclusions arrived at, as the basis for this view were not available for scrutiny.

However, it seems likely that a key consideration favouring such consoli-dation was the cushion it offered against the ongoing financial travails of the legacy-media sectors in Ireland. This was perhaps most overt in the case of the one merger that actually was subjected to a full investigation: that of the Celtic Media Group by INM. Ironically, in terms of market capitalization, it was the smallest of the mergers. The decision to refer it to the BAI for a full investigation was almost certainly informed by a political recognition that any further Denis O'Brien-related concentration of media ownership had at least to be seen to be properly scrutinized. Yet, from the Celtic Media Group's perspective, even though the merger did not ultimately proceed, it was about survival: there was clearly some security to be found in becoming part of a larger media entity. Though this was particularly the case with regard to Irish print titles, it also held true for broadcast entities. For the department of communications and the BAI, the question became whether media pluralism was best served by preventing further concentration (which risked seeing smaller media firms go out of business) or by allowing firms to improve their viability by becoming part of larger media entities even if it meant a further concentration of editorial power. In this regard it was notable that November 2016 saw the BAI emphasize 'sectoral sustainability' above other objectives in outlining the development of a new long-term strategy statement.

The one media entity that could not entertain such consolidation options was RTÉ. By 2016, the institution that had constituted the core of Irish broadcasting for nearly a century was confronted with the question of not just how but whether it could survive in a market increasingly colonized by channels replete with local content and supported by hugely capitalized inter-national parents. This context may have influenced the choice of successor to Noel Curran after he announced his decision to step down as DG with effect from May 2016. Dee Forbes was not only the first woman but the first person from outside RTÉ to occupy the role for half a century. In contrast to her immediate predecessors, whose early experience lay in programme-making, Forbes came from a sales-and-distribution background and brought to RTÉ several decades of international experience working for Ted Turner in the US and at Discovery Networks, where she had been president of the Western European division.

The context she inherited could hardly have been more fraught. A combination of increased expenditure associated with 1916 centenary celebrations (most notably the €6 million drama centrepiece broadcast in spring 2016, *Rebellion*) and an unexpected decline in commercial revenues as large UK-based advertisers curtailed their spending in the wake of the UK's Brexit decision, left RTÉ looking at a potential 2016 deficit of €20 million. Furthermore, Forbes' early hope that RTÉ could work with its newly recapitalized 'frenemies' seemed optimistic, as RTÉ lost several senior staff to Eir and TV3 in summer 2016. There was also a marked ramping up of competition for sports rights, facilitated by the failure of every minister for communications since 2008 to expand the 2003 list of events designated as free-to-air. Even where events were protected, the terms on which RTÉ accessed them had changed. Ironically, it was under Dee Forbes that Discovery Networks had, in June 2015, acquired the rights to the 2018 to 2024 summer and winter Olympic Games, forcing RTÉ to – in the main successfully – negotiate for access to coverage. Yet it was in rugby where the deep pockets of Liberty Global and Eir made the most profound impact. In November 2015, TV3, having just secured its largest-ever ratings based on Irish matches in the summer 2015 Rugby World Cup, spent a reported €20 million to outbid RTÉ for the rights to the rugby Six Nations Championship from 2018 to 2021. Less than a year later, in July 2016, Eir secured Irish rights to all games for the 2019 Rugby World Cup in Japan and, in November of the same year, exclusively transmitted a never-before-achieved Irish rugby team victory over New Zealand. Not only was RTÉ unable to compete for these events, but by early 2016 the parlous state of its finances saw the station sub-licence some of its existing sports rights to raise revenues. Thus TV3 unexpectedly acquired rights to twenty-two of the fifty-one games played during the football European Championship in summer 2016 (including the final, which it simulcast with RTÉ).

If RTÉ could not retain rights to the events featuring the national team, where was it to find content that mattered to a local audience? On retiring in 2016, Noel Curran had described foreign acquisitions as a 'sunset industry': as access to linear and on-demand foreign channels proliferated, so did ease of access to overseas drama content. In theory this made local content – and especially drama – more valuable than ever, if only as a means of distinguishing RTÉ from its competitors. Yet, how could this be funded given the ongoing reluctance of the state to countenance either an increase in the licence fee or its replacement by a harder-to-evade household broadcasting charge?

In April 2015, Alex White had confirmed that the household broadcasting charge would not be introduced within the lifetime of the government, until 'public understanding and support' for such a move were built up. He did, however, signal his intention to introduce legislation forcing Sky and UPC to hand over their customer data to An Post to help it combat licence-fee evasion; by October 2015, with an election looming and heightened RTÉ/state tension over broadcast coverage of water-charge protests, even that proposal was quietly dropped. Though Denis Naughten would argue that there was 'huge merit' in the principle of a household broadcasting charge in June 2016, he also argued that the new government simply didn't have the Dáil numbers to get it through (*IT*, 14 Oct. 2016).

Caught between a rock and hard place, by September 2016, Forbes made it clear that she was assessing the future viability of 'everything' RTÉ did. In 2013, RTÉ had dismissed the idea that increased reliance on independent commissions would lead to cost savings. In November 2016, the station announced that it was outsourcing all children's television production (then worth about €11 million) to the independent sector, raising the prospect that other areas might follow. In March 2017, in an address to staff Forbes announced the need to cut employee numbers by 10 per cent. No less wedded than her predecessor to the importance of developing RTÉ's digital-production capacity, she also announced that, decades after it was first mooted, RTÉ would sell some nine acres of the Montrose site to raise €75 million for infrastructure and capital development. Yet such one-off injections of funds were short-term solutions and could not address the larger issue of how to sustain a key element of the Irish public sphere into the future.

Coda: *Quo vadis?*

The communications media in Ireland, as elsewhere, have experienced and will continue to experience what is, at times, traumatic change. The long view offered by this volume also demonstrates that in some respects, legacy media forms have been remarkably long-lived, notwithstanding the profound changes visited upon the twentieth century. The basic structure of the commercial newspaper industry as established in the second half of the nineteenth century persisted more or less unchanged until the early years of the twenty-first. Broadcasting too, though augmented by television from the 1950s, operated around a range of state-owned institutions until the close of the twentieth century.

Nonetheless, the remaining stalwarts of twentieth-century Irish media – RTÉ, the *Irish Times*, the *Irish Independent* and many surviving local newspapers – are experiencing irrevocable change. RTÉ's position is precarious. At the *Irish Times*, though radical cost-cutting stabilized the initial post-recession losses, the paper's average revenues of between €80 million and €90 million since 2010 are well below the nearly €140 million it recorded during the boom. That paper's 2015 annual report described print as a 'challenging sector which is subject to structural decline'. This seems no less true of Independent News and Media: its now reduced profitability owes more to distribution and commercial printing activities than to advertising and circulation revenues, which – small absolute increases in digital revenues notwithstanding – are still fragile. Digital revenues are increasing, but not substantially enough to plug the gap created by print-circulation and advertising weakness.

This volume has pointed to a long relationship between political and media institutions, which, though characterized by occasional flashpoints, has for the most part been one of mutual, if occasionally grudging, accommodation. This is in part because RTÉ, which is still the single most influential Irish media institution, though not a creature of the state, is directly owned by it. However, Irish media more generally can be understood as an element of the institutional framework constituting Ireland's particular performance of liberal democracy. Eric Louw (2010) has argued that the coincidence of the emergence of a mass media in the nineteenth and twentieth centuries with the spread of suffrage was driven less by an emancipatory desire to educate the newly politicized masses and more by a desire to use mass media organs to ensure that the newly enfranchised masses (or 'the mob') did not upset the political and social status quo. For most of the twentieth century, the concentration of media power in the hands of well-capitalized private or state-owned institutions narrowed access to these means of expression and, by extension, to social and political influence. What is more, as Strömböck (2008) has proposed, the increasing reliance of politics on media outlets to reach the public has seen a concomitant increase in the influence of mass-media logics upon political performers and political content. As a concept 'mediatization' suggests that the mass media and the political sphere became inextricably intertwined to form the core of twentieth-century liberal democracies.

Yet, even before the advent of social media, the cost barriers to accessing technologies of media production and, subsequently, distribution, were falling. With the rise of social media, these costs were virtually reduced to zero: the launch of Facebook Live, for example, in April 2016 meant that anyone with a smartphone and Facebook could live-stream audiovisual content from anywhere and to anywhere on the planet with internet access. Such changes both reflect and facilitate the possibility of a change in traditional institutions of power based around what Castells (2007) has characterized as a 'new form of socialized communication: mass self-communication'. Though the bulk of social-media communications are of a personal character, Castells has argued that mass self-communication also offers 'an extraordinary medium for social movements and rebellious individuals to build their autonomy and confront the institutions of society in their own terms and around their own projects', with a view to modifying existing power relations.

Though Castells' occasional identification of social media with liberatory potential is at times somewhat breathless, his thesis seems at least partially underwritten by events in the decade after he wrote the sentence cited above, albeit perhaps not quite in the progressive direction he might have hoped for. The rise of populist politics in the 2010s, whereby the 'outsider' (or at least those who can successfully pass themselves off as such) triumphed across Western liberal democracies, clearly took established political elites by surprise. And the rise of Trump in the US, Brexit in the UK and far-right politics across continental Europe was at least in part based on the exploitation of alternative networks of communication – Trump's use of Twitter, hate groups on Facebook – which bypassed the communicative bottleneck of legacy media. In other words, 'the mob' found a new means through which to express itself.

This has given rise to a kind of moral panic regarding social media among members of the Irish political class, who call for regulation of social media because 'the modern mob just can't handle the power they have unwisely been granted' (Titley, 2013). Irish society has long been used to the idea that the public sphere is a space where freedom of expression can be regulated, either through direct ownership of media outlets (RTÉ and the Irish print media), or through direct (the BAI and the department of communications) or indirect (the Defamation Act and the press council) regulation. Though social media do not quite constitute the Wild West of some imaginings (Irish defamation law applies equally to online comment as it does to print or broadcast content), it clearly creates previously unavailable conduits of expression. As Titley notes, the kind of personalized invective aimed at Irish politicians through social media may also reflect a context wherein 'the mob' has repeatedly been informed by successive governments that 'there is no alternative' to austerity policies. If there's no alternative then the space for debate shuts down and abuse emerges in its stead.

Instances of established institutions struggling to adapt to (if not adopt) new media logics proliferated after 2008. The debacle over the Seán Gallagher tweet on *Frontline* came a year after an earlier incident connected to the passing of broadcaster Gerry Ryan, news of whose death was inadvertently broken by some of his colleagues on Twitter before it had been officially confirmed. In 2014, after RTÉ's swift apology in the wake of the 'homophobia row' on *The Saturday Show*, the subject appeared to have been shut down. Yet, weeks later, Panti Bliss was invited to give an address – a 'noble call' – on the subject of homophobia before an audience in Dublin's Abbey Theatre. The film record of the event was uploaded to YouTube, where it went viral, ultimately reaching a geographically and numerically far wider audience than had seen the original RTÉ broadcast.

But perhaps the most far-reaching illustration of the extent to which new media could circumvent the limitations of old was associated with the water-charge protests. Beginning in 2012, the protest engendered what was arguably the longest active civil protest movement in the history of the state. Though politicians would accuse broadcasters of feeding the movement with the oxygen of publicity, protest groups argued that, if anything, their actions were either under-reported or dismissed. The longevity of the movement probably owed more to its exploitation of the alternative public sphere offered by outlets such as Facebook and YouTube. Tens of thousands of examples of water-charge-protest 'sousveillance' videos appeared on YouTube throughout 2013 and 2014. Even if more established media outlets had entirely ignored the protests, there was ample material available offering some public exposure to them. Political frustration aimed at RTÉ in this regard may well have been displaced anger at the harder-to-regulate nature of such social media.

Such conclusions must be tentative because the kind of in-depth research needed to validate them is still relatively nascent in Ireland. The academic study of a medium disseminating diffuse fragments to even more diffuse audiences creates new challenges even as it creates new opportunities for institutions that promise to make sense of it all.

The narrative of Storyful is instructive in this regard. Having left RTÉ in January 2010, journalist Mark Little established this company as a social-media news agency that sifted through user-generated content for newsworthy material and, having verified its authenticity, acted as an intermediary between the content creator and mainstream media organizations. Thus legacy media could safely include, for example, YouTube content in their own output. In a telling reflection of, literally, changing media values, Storyful, which produced no content of its own, was in December 2013 sold to News Corp for €18 million, making it more valuable than all but the largest Irish legacy media combinations.

Yet even if Storyful suggested that the relationship between legacy and online media was under negotiation, the 2016 results from the annual Digital News Report survey carried out for Reuters by the DCU Institute for Future Journalism and Media (Fujo) suggested that legacy media continued to exert significant influence upon Irish audiences. Though online sources vied for top spot, with 70 per cent of respondents identifying them as a weekly source of news, television remained the leading source, with 73 per cent. Against that more Irish audiences resorted to both online sources and specifically social media (52 per cent) for news each week than the 47 per cent who looked to radio or the 45 per cent who read a printed newspaper.

Among legacy media, RTÉ television and radio remained popular: 36 per cent (more than for any other brand) cited RTÉ as their main source of news, as compared with the 6 per cent who cited the *Irish Independent* or Today FM. Online, the fragmentation of audiences diluted RTÉ's impact, but as of 2016 it remained the leading brand – 42 per cent of respondents used RTÉ's online news each week – followed by another legacy outlet, *Independent.ie,* which 36 per cent of respondents used. Net-native outlets were biting at the heels of the front-runners: 34 per cent cite *thejournal.ie* as their main online news source. (Indeed, though well behind established Irish brands online, it was striking that a number of not just UK-based sites, but also US sites such as *Huffington Post* or the *New York Times*, featured among the top twenty most followed online news source in Ireland.)

The numbers identifying the online space as a source of news raised the question of whether legacy media might be supplanted by alternatives. That 59 per cent of respondents went straight to trusted brands suggested that, notwithstanding a decline in trust in the media as whole, most people continued to see some value in the story selection and interpretative frameworks used by recognized news brands.

Furthermore, if a substantial minority (31 per cent) of Irish respondents to the Reuters study simply 'happened upon' news through social media, it is important to acknowledge that most of this news was constituted by links to legacy media content. The leading social-media entity Facebook, for example, is not in and of itself a producer of news but a conduit between media outlets and Facebook users. Nonetheless, Facebook's influence in bringing users to individual stories changed the manner in which audiences accessed content from legacy outlets (while conferring considerable market *and* opinion-forming power on a company that, while hardly a disinterested observer of world events, operates in a largely unregulated fashion).

The Facebook business model relied on stratifying its user base so as to sell users on to advertisers. That model meant Facebook had to constantly survey user preferences to incentivize their ongoing presence, ensuring a constant flow of material addressing their particular interests. As DeVito (2016) pointed out, this was achieved by a series of constantly tweaked algorithms that applied reasonably clear – if unsurprising – criteria for selecting stories for the Facebook feeds of individuals. Friendship with a poster and semantic correlations between their posts and your own previous status updates (i.e. contributions to Facebook) combined with novelty as key determinants. The concern raised here was that of the so-called 'filter-bubble' effect, which ensures that we are more likely to encounter content reflecting our previously expressed preferences, limiting users to material that confirms their existing prejudices. The question this raised was where such editorial power should lie – with a traditional editor or the algorithm – and at what point should national governments, or international alliances such as the European Union, tackle the regulatory issues, and the threat to legacy media posed by the morphing of digital platforms into digital publishing operations?

Regardless of such developments, it seemed clear that, notwithstanding the struggle between legacy and online media institutions, they were nonetheless likely to become increasingly mutually interdependent. Social media needed content; legacy need audiences. The questions for the coming decade are how that relationship will resolve, what parts of it will be successfully monetized, and to what extent that resolution will serve the interests of society at large.

Bibliography

Adams, Michael. 1968. *Censorship: the Irish experience.* Tuscaloosa: University of Alabama Press.

Ahearne, Alan. 2007. 'The lie of the land', *Irish Times,* 28 Sept.

Akenson, Donal Harman. 1994. *Narrative.* Vol. 1 of *Conor: a biography of Conor Cruise O'Brien.* Montreal: McGill-Queen's University Press.

Amárach Consulting. 2000. 'Technology futures'. Dublin: Amárach.

——. 2003. 'Consumer TrendWatch Q2 2003'. Dublin: ComReg.

——. 2005. 'ComReg trends report Q1 2005'. Dublin: ComReg.

——. 2014. 'The second UPC report on Ireland's digital future'. Dublin: UPC.

——. 2016. 'Virgin Media digital insights report'. Dublin: Virgin Media.

Andrews, Ann. 2014. *Newspapers and newsmakers: the Dublin nationalist press in the mid-nineteenth century.* Liverpool: University Press.

Arnold, Bruce. 1984. *What kind of country: modern Irish politics, 1968–1983.* London: Jonathan Cape.

Artero, Juan Pablo, Roderick Flynn and Damian Guzek. Forthcoming. 'Media concentration and the rise of multinational companies'. In *The handbook of European Communication History,* ed. Paschal Preston, Klaus Arnold and Suzanne Kinnebrock. London: Wiley-Blackwell.

BAI. 2012. 'Ownership and control policy'. Dublin: BAI.

——. 2013. 'Five-year review of public funding: authority recommendations'. Dublin: BAI.

Baker, Stephen. 2005. 'The alternative press in Northern Ireland and the political process', *Journalism Studies,* 6(3): 375–86

Barclay, Andy. 1993. 'Final death notice for the *Sunday News*', *Irish Times,* 25 Mar.

Barrett, Sean D. 2000. *Competitiveness and contestability in the Irish media sector,* Trinity Economic Paper No. 3. Dublin: Department of Economics, Trinity College.

Becker, Carlo. 2016. 'After the transition', *New Left Review,* 99: 55–6.

Beesley, Arthur. 2008. 'O'Reilly v. O'Brien: the battle for a media empire', *Irish Times,* 12 Apr.

Behaviour and Attitudes. 2015. *Eir connected living survey 2015.* Dublin: Eir.

Bell, Emily and Owen Taylor. 2017. 'The platform press: how Silicon Valley re-engineered journalism'. New York: Tow Center for Digital Journalism.

Bell, J. Bowyer. 1969. 'Ireland and the Spanish Civil War, 1936–1939', *Studia Hibernica,* 9: 137–63.

Bell, Martin. 1973. 'Reporting from Ulster'. In *A second listener anthology,* ed. Karl Miller, 370–2. London: BBC.

Blanshard, Paul. 1953. *The Irish and Catholic power: an American interpretation.* Boston: Beacon Press.

Bourke, S. 2008. 'Ethical trends and issues in Irish journalism'. MA, Dublin City University.

Boyle, Kevin and Marie McGonagle. 1988. *Press freedom and libel.* Dublin: National Newspapers of Ireland.

——. 1995. *Media accountability: the readers' representative in Irish newspapers.* Dublin: National Newspapers of Ireland.

Brady, Conor. 2010. 'Did the media fail to sound alarm bells before the financial crisis?', *Irish Times*, 6 Mar.

——. 2012. 'Future must be secured for serious new media', *Irish Times*, 8 Feb.

Brodie, Malcolm. 1995. *The Tele: a history of the Belfast Telegraph.* Belfast: Blackstaff Press.

Brody, Annabel. 2012. 'Media accountability in the twenty-first century'. PhD, National University of Ireland, Galway.

Brown, Terence. 1981. *Ireland: a cultural and social history.* London: Fontana.

——. 2015. *The Irish Times: 150 years of influence.* London: Bloomsbury.

Browne, Noel. 1986. *Against the tide.* Dublin: Gill and Macmillan.

Browne, Stephen. 1937. *The press in Ireland: a survey and a guide.* Repr., New York: Lemma, 1971.

Browne, Vincent. 2007. 'Julia Kushnir case issues', *Irish Times*, 14 Nov.

Bundock, Clement J. 1957. *The National Union of Journalists: a jubilee history, 1907–1957.* Oxford: University Press.

Burke, Ray. 2005. *Press delete: the decline and fall of the 'Irish Press'.* Dublin: Currach Press.

Butler, David. 1995. *The trouble with reporting Northern Ireland.* Aldershot: Avebury Press.

Byrne, Gay. 1972. *To whom it concerns: ten years of the Late Late Show.* Dublin: Torc Books.

Callanan, Frank. 1996. *T.M. Healy.* Cork University Press.

Carolan, Mary. 2008. 'Woman wins €90,000 damages in privacy case', *Irish Times*, 19 July.

Castells, Manuel. 2007. 'Communication, power and counterpower in the network society', in *International Journal of Communication*, 1, 238–66.

Cathcart, Rex. 1972. 'TV coverage on Northern Ireland', *Index on Censorship*, 1(1): 15–32.

——. 1974. 'BBC NI: 50 years old', *The Listener*, 2372: 322–4.

——. 1984. *That most contrary region: the BBC in Northern Ireland.* Belfast: Blackstaff Press.

——. 1986. 'Mass media in twentieth-century Ireland'. In *Ireland, 1921–84*, vii: *a new history of Ireland*, ed. J.R. Hill, 671–710. Oxford: The University Press.

Cawley, Anthony. 2012. 'Sharing the pain or shouldering the burden? News-media framing of the public sector and the private sector in Ireland during the economic crisis, 2008–2010', *Journalism Studies*, 13 (4): 600–15.

——. 2016. 'Johnston Press and the crisis in Ireland's local newspaper industry, 2005–2014', *Journalism*, advance online publication: https://doi.org/10.1177/1464884916648092.

Chadwick, Andrew. 2013. *The hybrid media system.* Oxford University Press.

Chubb, Basil. 1972. 'Media and the state', *Irish Times*, 31 Aug. 1972.

——. 1974. *Cabinet government in Ireland*. Dublin: Institute of Public Administration.

——. 1984. 'The political role of the media in contemporary Ireland'. In *Communications and community in Ireland*, ed. Brian Farrell, 75–86. Cork: Mercier Press.

Citizens for Better Broadcasting. 1978. *Aspects of RTÉ television broadcasting*. Dublin: Citizens for Better Broadcasting.

Clarke, Paddy. 1986. *'Dublin calling': 2RN and the birth of Irish radio*. Dublin: Raidió Teilifís Éireann.

CNI. 1996. 'Report', PN 284 1. Dublin: Stationery Office.

Coleman, Nicola. 2016. 'Organise, win, and then tell your story', *The Journalist*, Oct.-Nov., 9.

Collins, John. 2006. 'Online advertising to reach €52m in 2007, meeting told', *Irish Times*, 19 Oct.

Collins, Stephen. 2016. 'Naughten considers funding print media', *Irish Times*, 14 Oct.

Commission on the Newspaper Industry. 1996. *Report: 2841*. Dublin: Government Publications.

Committee on Industrial Progress. 1970. 'Report on paper, paper products, printing and publishing industry', Prl 1356. Dublin: Stationery Office.

Committee on Irish Language Attitudes. 1975. 'Report'. Dublin: Stationery Office.

Competition and Mergers Review Group. 2000. 'Final report', PN 8487. Dublin: Department of Enterprise, Trade and Employment.

Competition Authority. 1992. 'Report of investigation of the proposal whereby Independent Newspapers plc would increase its shareholding in the Tribune Group from 29.9% to a possible 53.09%'. Dublin: Stationery Office.

——. 1995. 'Interim report of study on the newspaper industry'. Dublin: Stationery Office.

ComReg. 2003. 'Key data for Irish communications market – September 2003'. Dublin: ComReg.

——. 2006. 'Irish communications market: key data report – March 2006'. Dublin: ComReg.

——. 2010. 'Irish communications market: key data report – Q1 2010'. Dublin: ComReg.

——. 2011. 'Irish communications market: key data report – Q3 2011'. Dublin: ComReg.

——. 2016. 'Irish communications market: key data report – Q4 2016'. Dublin: ComReg.

Coogan, Tim Pat. 1993. 'Dev and the Irish press: from national crusade to nest-egg?' *Irish Times*, 29 Sept.

Cooney, John. 1999. *John Charles McQuaid: ruler of Catholic Ireland*. Dublin: O'Brien Press.

Corcoran, Farrel. 1996. 'Media, children and RTÉ', *Irish Communications Review*, 6: 83–9.
——. 2009. RTÉ and the globalization of Irish television. Bristol: Intellect.
Coulter, Carol. 2009. 'Privacy bill does not pose threat to good journalism, says minister', *Irish Times*, 11 May.
Courts Service. 2007. Judge Mahon -v- Keena & Anor, Neutral Citation: [2007] IEHC 348.
——. 2008. Herrity -v- Associated Newspapers (Ireland) Ltd., Neutral Citation: [2008] IEHC 249.
——. 2009a. Mahon Tribunal -v- Keena & Anor. Neutral Citation: [2009] IESC 64.
——. 2009b. Mahon Tribunal -v- Keena & Anor. Neutral Citation: [2009] IESC 78.
Crowe Horwath. 2013. 'Five-year review of public-service broadcasting'. Dublin: BAI.
Cullen, Frank. 2009. Letter to Dermot Ahern, minister for justice, 18 May. National Newspapers of Ireland Archives.
——. 2013. 'Watchdog must address RTÉ's commercial agenda', *Irish Times*, 7 May.
Cullen, Paul. 1991. 'An Irish World Service: the story of Ireland's shortwave broadcasting station'. MA, Dublin City University.
——. 2007. 'Lawlors seek copy of death coverage inquiry', *Irish Times*, 8 Nov. 2007.
Curtis, Liz. 1984. *Ireland – the propaganda war*. London: Pluto.
Darby, John. 1983. *Dressed to kill: cartoonists and the Northern Ireland conflict*. Belfast: Appletree Press.
Davis, E.E. and R. Sinnott. 1979. 'Attitudes in the Republic of Ireland relevant to the Northern Ireland problem, Vol. 1', ESRI paper No. 97. Dublin: Economic and Social Research Institute.
Dawson, Kevin. 2009. 'Why TV3's investors really want a bigger slice of your licence fee', *Irish Times*, 9 July.
De Bréadún, Deaglán. 2002. 'Haughey and RTÉ clash over editing', *Irish Times*, 1 Jan.
Dekavalla, Marina and Kevin Rafter. 2016. 'The construction of a "historical moment": Queen Elizabeth's 2011 visit to Ireland in Irish and British newspapers', *Journalism*, 17 (2): 227–43.
Department of Arts, Culture and the Gaeltacht. 1995. 'Active or passive? Broadcasting in the future tense: green paper on broadcasting'. Dublin: Stationery Office.
Devane, R.S. 1927b. 'Suggested tariffs on imported newspapers and magazines', *Studies*, 16: 545–63.
De Vere White, Terence. 1967. 'Social life in Ireland 1927–37'. In *The years of the great test, 1926–39*, ed. Francis MacManus. Cork: Mercier Press.
Devenport, Mark. 2000. *Flash frames: twelve years reporting Belfast*. Belfast: Blackstaff Press.
DeVito, Michael A. 2016. 'From editors to algorithms', *Digital Journalism*, advance online publication: http://www.tandfonline.com/doi/full/10.1080/21670811. 2016.1178592.
Doolan, Lelia, Jack Dowling and Bob Quinn. 1969. *Sit down and be counted: the cultural evolution of a television station*. Dublin: Wellington.

Doyle, Gillian. 2000. 'The economics of monomedia and cross-media expansion: a study of the case favouring deregulation of TV and newspaper ownership in the UK', *Journal of Cultural Economics*, 24: 1–26.

Duggan, Keith. 2010. 'Forget issues, our nation has been Sky-ified', *Irish Times*, 22 May.

Duncan, Pamela. 2011. 'Review of RTÉ radio funding urged', *Irish Times*, 14 Sept.

Dwyer, T. Ryle. 2005. 'Why the housing boom could collapse like a ton of bricks', *Irish Examiner*, 5 May.

ECHR. 2014. Case No. 178 Oct. 2014, Keena and Kennedy v. Ireland (Dec.) – 29804/10 Decision 30.9.2014 [Section V].

Eir. 2013a. 'Home sentiment survey phase II, April 2013'. Eir: Dublin.

——. 2013b. 'Home sentiment survey phase III, September 2013'. Eir: Dublin.

——. 2014a. 'Home sentiment survey phase IV, February 2014'. Eir: Dublin.

——. 2014b. 'Home sentiment survey phase V, September 2014'. Eir: Dublin.

Eurobarometer. 2011. 'European Barometer special 359: attitudes on data protection and electronic identity in the European Union'. Brussels: DG Communications.

Fahy, Declan, Mark O'Brien and Valerio Poti. 2010. 'From boom to bust: a post-Celtic Tiger analysis of the norms, values, and roles of Irish financial journalists', *Irish Communications Review*, 12: 5–20.

Fanning, Ronan. 1983. *The four-leafed shamrock: electoral politics and the national imagination in independent Ireland*. Dublin: National University of Ireland.

Farrell, David. 1991. 'Public broadcasting in a new state: the debate over the foundation of Irish radio, 1922–26', Manchester Papers in Politics, 7/91. Manchester: Victoria University.

Fawcett, L. 2002. 'Why peace journalism isn't news', *Journalism Studies*, 3. http://dx.doi.org/10.1080/14616700220129982

Feeney, Brian. 1997. 'The peace process: who defines the news – the media or the government press offices?' In *Media in Ireland: the search for diversity*, ed. Declan Kiberd, 41–59. Dublin: Open Air.

Feeney, Peter. 1985. *The television treatment of the subject of unemployment*. Dublin: RTÉ.

Felle, T. 2012. 'From boom to bust: Irish local newspapers post the Celtic Tiger'. In *What do we mean by local? Grass roots journalism: its death and rebirth*, ed. J. Mair, N. Fowler and I. Reeves, 41–50. Bury St Edmonds: Abramis.

Fianna Fáil/Progressive Democrats. 2002. 'Programme for government'. Dublin: Fianna Fáil.

Finn, Vincent. 1993. 'Thirty years a'growing: the past, the present and the future of Irish broadcasting', *Irish Communications Review*, 3: 73–8.

Fisher, Desmond. 1978. *Broadcasting in Ireland*. London: Routledge and Kegan Paul.

Fleck, Tony. 1995. 'Thatcher, the IBA and "Death on the Rock"', *Irish Communications Review*, 5: 1–11.

Fleming, Lionel. 1965. *Head or harp*. London: Barrie & Rockliff.

Flood Tribunal (Tribunal of Inquiry into Certain Planning Matters and Payments).

2002. *The second interim report of the Tribunal of Inquiry into Certain Planning Matters and Payments.* Dublin: Government Publications.

Flynn, Gerard. 1983. 'Newspapers in crisis', *Business and Finance,* 23 Sept.

Flynn, Roderick. 2014. 'How important shall public service media be in the European digital media age?' In *Public Value Report 2013/14,* ed. Konrad Mitschka, 59–63. Vienna: ORF.

——. 2015. 'Public service broadcasting beyond public service broadcasters', *International Journal of Digital Television,* 6(2): 125–44.

Foley, Michael. 1994. 'Atlantic helps shatter BBC radio leadership', *Irish Times,* 25 Oct.

Fox, Carl, Steven Knowlton, Aine Maguire and Brian Trench. 2009. 'Accuracy in Irish newspapers: a report for the Press Council of Ireland', https://www.academia.edu/28341255/Accuracy_in_Irish_newspapers_a_report_for_the_Press_Council_of_Ireland.

French, Nora. 2007. 'Journalism education in Ireland', *Irish Communications Review,* 10: 41–9.

Gailey, Andrew. 1995. *Crying in the wilderness: Jack Sayers: a liberal editor in Ulster, 1939–69.* Belfast: Institute of Irish Studies, Queen's University, Belfast.

Geary, Frank. 1955. 'I remember', *Irish Independent,* 3 Jan.

Gibbons, Luke. 1984. 'From kitchen sink to soap'. In *Television and Irish society: 21 years of Irish television,* ed. Martin McLoone and John MacMahon, 21–53. Dublin: RTÉ/IFI.

——. 1996. *Transformations in Irish culture.* Cork University Press.

Goan, Cathal. 2009. 'Letters to the editor', *Irish Times,* 2 May.

Gorham, Maurice. 1967. *Forty years of Irish broadcasting.* Dublin: Talbot Press for RTÉ.

Gray, Tony. 1991. *Mr Smyllie, sir.* Dublin: Gill and Macmillan.

Hackett, Robert A. 2006. 'Is peace journalism possible? Three frameworks for assessing structure and agency in news media', *Communications and Conflict Online,* 5(2): 1–13.

Hancock, Ciaran. 2011. 'INM rejects O'Brien's claims that he is being "punished" by company's coverage', *Irish Times,* 17 Nov.

Hanitzsch, Thomas. 2004. 'Journalists as peacekeeping force? Peace journalism and mass communications theory', *Journalism Studies,* 5 (2004).

——. 2007. 'Situating peace journalism in journalism studies: a critical appraisal', *Conflict & Communication Online,* 6 (2).

Hannigan, David. 1993. 'The *Kilkenny People* and the censors, 1939–45'. MA, Dublin City University.

Hayley, Barbara and Enda McKay (eds). 1987. *Three hundred years of Irish periodicals.* Dublin: Association of Irish Learned Journals/Lilliput Press.

Hazelkorn, Ellen. 1996. 'New technologies and changing work practices in the media', *Irish Communications Review,* 6: 28–39.

Henderson, R.B. 1984. *A musing on the lighter side of Ulster Television and its first 25 years.* Belfast: Universities Press for UTV.

Hickie, Aileen. 1994. 'The history and structure of *An Phoblacht*'. Unpublished term paper, Dublin City University.

Horgan, John. 1984a. 'The press: credibility and accountability', *The Crane Bag*, 8 (2): 64–75.

——. 1984b. 'State policy and the press', *The Crane Bag*, 8 (2): 51–61.

——. 1986. 'The provincial papers of Ireland: ownership and control and the representation of "community". In *Is the Irish press independent?*, ed. D. Bell. Dublin: Media Association of Ireland.

——. 1993a. 'Government, propaganda and the Irish News Agency', *Irish Communications Review*, 3: 31–43.

——. 1993b. 'Over the sea from Skye: cross-channel competition and the Irish media market', *Teaglaim*, 1: 35–9.

——. 1995. 'Saving us from ourselves: contraception, censorship and the "evil literature" controversy of 1926', *Irish Communications Review*, 5: 61–7.

——. 1997. *Sean Lemass: the pragmatic patriot.* Dublin: Gill and Macmillan.

——. 1999. 'The media and the enemies of truth'. In *New century, new society: Christian perspectives*, ed. Dermot Lane, 93–102. Dublin: Columba Press.

——. 2004. *Broadcasting and public life: RTÉ news and current affairs, 1926–1997.* Dublin: Four Courts Press.

——. 2008. 'Press Council's suppression of dissenting voices forced me to quit: the Press Council should be the last body to suppress minority or dissenting opinions for the sake of collegiality', *Irish Times*, 12 May.

Hourigan, Niamh. 1996. 'Audience identification and Raidió na Gaeltachta', *Irish Communications Review*, 6: 1–6.

Humphreys, Joe. 2011. 'Missionary movement must take a hard look at its modern self', *Irish Times*, 30 Aug.

Inglis, Brian. 1954. *The freedom of the press in Ireland, 1784–1841.* London: Faber.

Integra. 2000. *Perception is power: social exclusion and the media.* Dublin: WRC Social and Economic Consultants.

Irish Times. 2015. 'Former editor says 'Irish Times' not compromised by property advertising', 27 Mar.

Kelly, Mary. 1984. 'Twenty years of current affairs on RTÉ'. In *Television and Irish society: 21 years of Irish television*, ed. Martin MacLoon and John MacMahon, 89–107. Dublin: RTÉ/IFI.

——. 1992. 'The media and national identity in Ireland'. In *Ireland and Poland: comparative perspectives*, ed. P. Clancy et al., 75–89. Dublin: Department of Sociology, University College, Dublin.

Kelly, Mary and Bill Rolston. 1995. 'Broadcasting in Ireland: issues of national identity and censorship'. In *Irish society: sociological perspective*, ed. P. Clancy et al., 563–92. Dublin: Institute of Public Administration.

Kelly, Morgan. 2006a. 'Irish house prices: gliding into the abyss', http://www.csn.ul.ie/~hugh/att/housing1.pdf, accessed 18 Sept. 2016.

——. 2006b. 'How the housing cornerstones of our economy could go into rapid free-fall', *Irish Times*, 28 Dec.

——. 2006c. 'Soft landing?', http://www.csn.ul.ie/~hugh/att/housing1.pdf, accessed 23 Apr. 2006.

——. 2009. 'The Irish credit bubble', Geary Discussion Papers Series. Dublin: University College Dublin.

Kelly, Olivia. 2008. 'RTÉ warned on public service content', *Irish Times*, 3 Mar.

Kenneally, Ian. 2008. *The paper wall: newspapers and propaganda in Ireland, 1919–1921.* Cork: Collins Press.

Kennedy, Dennis. 1988. *The widening gulf: Northern attitudes to the independent Irish state, 1919–49.* Belfast: Blackstaff Press.

Kennedy, Geraldine. 2014. 'A cold, calculated decision to step outside the law', *Irish Times*, 25 Oct. 2014.

Kenny, Colum. 1998. 'The politicians, the promises and the mystery surrounding Radio Bonanza', *Sunday Independent*, 21 June.

Kenny, Robert and Tim Suiter. 2015. *Ownership and control of media businesses in Ireland.* Dublin, BAI.

Keogh, Dermot. 1994. *Twentieth-century Ireland.* Dublin: Gill and Macmillan.

——. 1995. 'Ireland and "Emergency Culture": between Civil War and normalcy, 1922–1961', *Ireland: A Journal of History and Society*, 1 (1): 21–33.

Kiberd, Declan, ed. 1997. *Media in Ireland: the search for diversity.* Dublin: Open Air.

Kilfeather, Frank. 1997. *Changing times: a life in journalism.* Dublin: Blackwater Press.

Kovarik, B. 2006. 'Peace journalism: media and the path to peace', *Journalism and Mass Communications Quarterly*, 83 (2): 444.

KRW Law/Doughty Street Chambers. 2016. 'Report on the concentration of media ownership in Ireland'. Brussels: GUE/NGL.

Kyle, Keith. 1969. 'The Ulster Emergency and BBC's impartiality', *The Listener*, 4: 297–9.

LANI (Library Association of Northern Ireland). 1987. *Northern Ireland newspapers, 1737–1987: a checklist with locations.* Belfast: Library Association.

Larkin, Felix. 2013. 'Double helix: two elites in politics and journalism in Ireland, 1870–1918'. In *Irish elites in the nineteenth century*, ed. Ciaran O'Neill, 125–36. Dublin: Four Courts Press.

——. 2014. 'Green shoots of the new journalism in the *Freeman's Journal*, 1877–1890'. In *Ireland and the New Journalism*, ed. K. Steele and M. de Nie. New York: Palgrave Macmillan.

Lee, Joseph (J.J.). 1989. *Ireland, 1912–1985: politics and society.* Cambridge University Press.

——. 1997. 'Democracy and public service broadcasting in Ireland'. In *Media in Ireland: the search for diversity*, ed. Declan Kiberd. Dublin: Open Air.

Legal Advisory Group on Defamation. 2002. *Report.* Dublin: Government Publications.

Legge, Marie-Louise. 1999. *Newspapers and nationalism: the Irish provincial press, 1850–1892.* Dublin: Four Courts Press.

Longford, the earl of, and Thomas O'Neill. 1970. *Eamon de Valera.* London: Hutchinson.

Louw, Eric. 2010. *The media and political process.* London: Sage.

Loyn, David. 2007. 'Good journalism or peace journalism?', *Conflict & Communication Online,* 6 (2), www.cco.regener-online.de, ISSN 1618–0747.

Lucey, Cornelius. 1937. 'The freedom of the press', *Irish Ecclesiastical Record,* 8: 589–99.

Mac Conghail. 1979. *Television in the eighties.* Review paper presented to RTÉ Authority seminar, 14/15 Dec.

MacDermott, Linda. 1995. 'The political censorship of the Irish broadcast media, 1960–1994'. MA, University of Notre Dame.

McGreevy, Ronan. 2011. 'RTÉ chief refuses to rule out dismissals', *Irish Times,* 23 Nov.

McGonagle, M. and A. Brody. 2013. 'Model for a cross-country comparative analysis of state media and communications regulatory bodies: the case of Ireland'. In *Media regulators in Europe: a cross-country comparative analysis,* ed. H. Sousa, W. Trutzschler, J. Fidalgo and M. Lameiras, 81–90. Braga, Portugal: Communication and Society Research Centre.

McGurk, Tom. 2000. 'Time for Irish TV for the North', *Sunday Business Post,* 11 June 2000.

McLoone, Martin and John MacMahon. 1984. *Television and Irish society: 21 years of Irish Television.* Dublin: RTÉ/IFI.

McLoone, Martin, ed. 1991. *Culture, identity and broadcasting in Ireland: local issues, global perspectives.* Belfast: Institute of Irish Studies.

——, ed. 1996. *Broadcasting in a divided community.* Belfast: Institute of Irish Studies.

Mercille, Julien. 2013. 'The role of the media in sustaining Ireland's housing bubble', *New Political Economy,* 19 (2): 282–301.

——. 2014. 'The role of the media in fiscal consolidation programmes: the case of Ireland', *Cambridge Journal of Economics,* 38: 281–300.

——. 2015. *The political economy and media coverage of the European economic crisis: the case of Ireland, 2008–2012.* London: Routledge.

M.H. Consultants. 1978. *A study of the evolution of concentration in the Irish publishing industry,* Evolution of Concentration and Competition Series, 14. Brussels: EU Commission.

Miller, David. 1994. *Don't mention the war: Northern Ireland, propaganda and the media.* London: Pluto.

Milotte, Mike. 2012. 'After the libel: how RTÉ investigations must change', *Irish Times,* 12 May.

Morash, Christopher. 2010. *A history of the media in Ireland.* Cambridge University Press.

Moriarty Tribunal (Tribunal of Inquiry into Payments to Politicians and Related Matters). 2011. *Report: Part II: Volume 1.* Dublin: Government Publications.

MRBI. 1983. '21st anniversary poll: the people of Ireland – a tribute'. Dublin: Market Research Bureau of Ireland.

Mulryan, Peter. 1988. *Radio Radio: the story of independent, local, community and pirate radio in Ireland.* Dublin: Borderline Press.

Munter, Robert LaVerne. 1967. *A history of the Irish newspaper, 1685–1760.* Cambridge University Press.

Newenham, Pamela. 2015. 'Facebook's Irish boss accentuates the positive', *Irish Times*, 9 Sept.

Nic Pháidín, Caoilfhionn. 1987. 'Na hIrsí Gaeilge'. In *300 years of Irish periodicals*, ed. Barbara Hayley and Enda McKay, 69–85. Dublin: Lilliput Press.

NNI. 1995. 'What the newspaper industry doesn't need from government'. Submission to the Commission on the Newspaper Industry. Dublin: NNI.

——. 1998b. 'Code of practice on privacy'. Dublin: NNI.

——. 1998. Fact sheet on the Freedom of Information Act 1997. Dublin: NNI.

O'Brien, Carl. 2008. 'Supreme court judge criticizes media', *Irish Times*, 26 Nov.

O'Brien, Conor Cruise. 1998. *Memoir: my life and themes.* Dublin: Poolbeg.

O'Brien, Denis. 1992. 'Independent local radio – an Irish success story', *Irish Communications Review*, 2: 54–8.

——. 2011. 'Depiction of me as an enemy of journalism undeserved', *Irish Times*, 15 Nov.

O'Brien, Mark. 2001. *De Valera, Fianna Fáil and the Irish Press: the truth in the news.* Dublin: Irish Academic Press.

——. 2008. *The* Irish Times: *a history.* Dublin: Four Courts Press.

O'Brien, Mark and Felix Larkin. 2014. *Periodicals and journalism in twentieth-century Ireland.* Dublin: Four Courts.

O'Brien, Mark and Kevin Rafter. 2012. *Independent Newspapers: a history.* Dublin: Four Courts Press.

Ó Cíosáin, Éamon. 1993. *An tÉireanneach, 1934–37: Nuachtán Soisíalach Gaeltachta.* Baile Atha Cliath: An Clochomhar.

O'Connell, Siobhan. 2009. 'TV3 argues for fair share of BCI programme funding', *Irish Times*, 25 June.

——. 2010. 'Newspapers step up campaign over RTÉ digital activities', *Irish Times*, 16 July.

O'Connor, Barbara. 1984. 'The presentation of women in Irish television drama'. In *Television and Irish society: 21 years of Irish television*, ed. Martin McLoone and John MacMahon, 123–33. Dublin: RTÉ/IFI.

O'Dell, Eoin. 2010. 'Defamation Act a welcome but imperfect reform for libel cases', *Irish Times*, 18 Jan.

O'Donnell, Donat [Conor Cruise O'Brien]. 1945a. 'The *Irish Independent*: a business idea', *The Bell*, 9 (5): 386–94.

——. 1945b. 'The Catholic press: a study in theopolitics', *The Bell*, 9 (5): 386–95.

Ó Drisceoil, Donal. 1996a. *Censorship in Ireland, 1939–45.* Cork University Press.

——. 1996b. 'Moral neutrality: censorship in Emergency Ireland', *History Ireland,* 4 (2): 46–50.

ODTR. 2002a. 'Irish communications market quarterly key data Sept. 2002'. Dublin: ODTR.

——. 2002b. 'Future delivery of broadband in Ireland'. Dublin: ODTR.

——. 2006. 'Irish communications market: key data report – March 2006'. Dublin: ODTR.

Ó Glaisne, Rísteárd. 1982. *Raidió na Gaeltachta.* Indreabhán: Cló Chois Fharraige.

O'Halloran, Marie. 2014. 'RTÉ got it wrong on Iona payout, says TD', *Irish Times,* 7 Feb.

O'Halpin, Eunan. 1999. *Defending Ireland: the Irish state and its enemies since 1922.* Oxford: University Press.

Ó hUanacháin, Micheál. 1980. 'The broadcasting dilemma', *Administration,* 28 (1): 33–71.

Oireachtas Éireann. 2016. 'Banking Inquiry: evidence', https://inquiries. oireachtas.ie/banking/hearings-evidence, accessed 8 Sept. 2016.

Oliver and Ohlbaum. 2013. 'Prospects for commercial digital terrestrial television in the Republic of Ireland'. BAI: Dublin.

O'Neill, Caithriona. 1984. 'Ownership and control in the provincial press'. BA, National Institute for Higher Education.

Oram, Hugh. 1983. *The newspaper book: a history of newspapers in Ireland, 1649–1983.* Dublin: MO Books.

O'Reilly, Emily. 1998. *Veronica Guerin: the life and death of a crime reporter.* London: Vintage.

O'Rorke, Andrew. 2006. 'Planned legislation not the answer to privacy concerns', *Irish Times,* 8 July.

O'Sullivan, John. 2005. 'Delivering Ireland: journalism's search for a role online', *Gazette: the International Journal for Communication Studies,* 67 (1): 45–68.

O'Sullivan, Timothy. 1984. *Fair and accurate? The press and the amendment.* Dublin: Veritas.

O'Toole, Fintan. 1990. *A Mass for Jesse James: a journey through the 1980s.* Dublin: Raven Arts.

O'Toole, James. 1998. *Newsplan: report of the Newsplan project in Ireland.* London and Dublin: British Library/National Library of Ireland.

PCI. 2007. 'Code of practice'.

Peleg, Samuel. 2006. 'Peace journalism through the lens of conflict theory: analysis and practice', *Conflict & Communication Online,* 5 (2), www.cco.regener-online.de, ISSN 1618–0747.

Phoenix, Eamon, ed. 1995. *A century of Northern life: the Irish news and 100 years of Ulster History 1890s–1990s.* Belfast: Ulster Historical Foundation.

Picard, Robert. 1983. 'Government intervention and press subsidies', *Editor and Publisher,* 13 June.

Preston, Paschal. 1995. *Democracy and communication in the new Europe.* Hampton, NJ: Hampton Press.

Rafter, Kevin, ed. 2011. *Irish journalism before independence: more disease than profession.* Manchester: University Press.

——. 2012. 'When the "Wild West" came to the local newspaper market'. In *What do we mean by local? Grass roots journalism: its death and rebirth,* ed. J. Mair, N. Fowler and I. Reeves, 34–40. Bury St Edmonds: Abramis.

Rafter, K. and S. Knowlton. 2013. '"Very shocking news": journalism and reporting on a politician's illness', *Journalism Studies,* 14 (3): 355–70.

Television Commission. 1959. 'Report'. Dublin: Stationery Office.

Red C. 2011. 'De-Coding Digital Trends Ireland 2011'.

Review Group on State Assets and Liabilities. 2011. *Report of the review group on state assets and liabilities.* Dublin: Government Publications.

Riggins, Stephen John Harold, *Ethnic minority media: an international perspective.* London: Sage.

Rose, Richard. 1971. *Governing without consensus: an Irish perspective.* London: Faber. RTÉ Authority minutes, RTÉ Archives.

Sanchez-Tarererno, Alfonso. 1993. *Media concentration in Europe: commercial enterprise and the public interest.* Dusseldorf: European Institute for the Media.

Savage, Robert. 1996. *Irish television: the political and social origins.* Cork University Press.

——. 2015. *The BBC's Irish troubles: television, conflict, and Northern Ireland.* Manchester University Press.

Schulz, Thilo. 1999. *Das Deutschlandbild der Irish Times 1933–45.* Frankfurt-am Main: Peter Lang.

Seanad Éireann, 20 Apr. 2010, Vol. 202, No. 1, Column 25: http://oireachtasdebates.oireachtas.ie/debates%20authoring/debateswebpack.nsf/takes/seanad2010042000006?opendocument.

Shearman, Hugh. 1987. *News Letter 1737–1987: a history of the oldest British daily newspaper.* Belfast: News Letter Publications.

Sheehan, Helena. 1987. *Irish television drama: a society and its stories.* Dublin: RTÉ.

Sheehy, Ian D. 2003. *Irish journalists and litterateurs in late Victorian London, c.1870–1910.* DPhil, University of Oxford.

Siapera, Eugene. 2012. *Understanding new media.* London: Sage.

Silke, Henry. 2015. 'Ideology, class, crisis and power: the role of the print media in the representation of economic crisis and political policy in Ireland (2007–2009)'. PhD, Dublin City University.

Slattery, Laura. 2009. 'Boxer pulls out of digital TV service', *Irish Times,* 21 Apr.

——. 2015a. 'Minister resists radio sector's call for fund', *Irish Times,* 29 Apr.

——. 2015b. 'INM plans to spend up to €100 million on media firms here and in UK', *Irish Times,* 6 June.

Slattery, Laura and Hugh Linehan. 2016. 'RTÉ faces cuts unless government acts on funding, outgoing DG warns', *Irish Times,* 14 Apr.

Smith, Anthony. 1972. 'Television Coverage of Northern Ireland', *Index on Censorship* 1 (2): 15–32.

Stokes Kennedy Crowley. 1984. 'Review of Radio Telefis Eireann: report to the minister for communications'. Dublin: Department of Communications.

Strömböck, Jesper. 2008. 'Four phases of mediatization: an analysis of the mediatization of politics', *International Journal of Press/Politics*, 13(3): 228–46.

Suiter, Jane. 2015. 'Declining trust in media is not healthy for democracy', *Irish Times*, 23 June.

Telecom Éireann. 1990. 'Annual report'. Dublin: Telecom Éireann.

Television Commission. 1959. 'Report', PR 5090, laid before both houses of Oireachtas 29 May.

Thinkhouse. 2015. 'Youth Culture Report Q2 2015'. [Place: Publisher?].

Titley, Gavin. 2013. 'Budgetjam! A communications intervention in the politico-economic crisis in Ireland', *Journalism*, 14 (2): 292–306.

Tóibín, Colm. 1990. *The trial of the generals: selected journalism, 1980–1990.* Dublin: Raven Arts Press.

Trench, Brian and Gary Quinn. 2003. 'Online news and changing modes of journalism', *Irish Communications Review*, 9 (1).

Trinity Mirror. 2000. 'Company press release on its disposal of *Belfast Telegraph*', 17 Mar.

Truetzschler, Wolfgang. 1991a. 'Foreign investment in the media in Ireland', *Irish Communications Review*, 1 (1): 1–4.

——. 1991b. 'Broadcasting law and broadcasting policy in Ireland', *Irish Communications Review*, 1 (1): 24–37.

UK CC (United Kingdom Competition Commission). 2000. 'Report', July.

UTV. 1999. 'Annual report'. Belfast: Ulster Television.

Walsh, Maurice. 2015. *Bitter freedom: Ireland in a revolutionary world, 1918–1923.* London, Faber.

Watson, Iarfhlaith. 1997. 'A history of Irish language broadcasting: national ideology, commercial interest and minority rights'. In *Media audiences in Ireland*, ed. Mary J. Kelly and Barbara O'Connor, 212–30. Dublin: UCD Press.

Williams, Raymond. 1973. 'The question of Ulster'. In *A second listener anthology*, ed. Karl Miller, 15–18. London: BBC.

Woodman, Kieran. 1985. *Media control in Ireland, 1923–1983.* Carbondale: Southern Illinois University Press.

Index